Thinking about Life

Paul S. Agutter • Denys N. Wheatley

Thinking about Life

The History and Philosophy
of Biology and Other Sciences

 Springer

Paul S. Agutter
Theoretical and Cell Biology
Consultancy
26 Castle Hill
Glossop, Derbyshire
SK13 7RR
United Kingdom

Denys N. Wheatley
Director, BioMedES
Leggat House, Whiterashes Road
Keithhall, Inverurie
AB51 OLX
United Kingdom

Cover pictures:
Top motif is a protein density image of a nucleopore.
Background picture is a pencil sketch called "Developing Senses".
Copyright Denys N. Wheatley

ISBN 978-1-4020-8865-0   e-ISBN 978-1-4020-8866-7

Library of Congress Control Number: 2008933269

Printed on acid-free paper

9 8 7 6 5 4 3 2 1

springer.com

Preface

Our previous book, *About Life*, concerned modern biology. We used our present-day understanding of cells to 'define' the living state, providing a basis for exploring several general-interest topics: the origin of life, extraterrestrial life, intelligence, and the possibility that humans are unique. The ideas we proposed in *About Life* were intended as starting-points for debate – we did not claim them as 'truth' – but the information on which they were based is currently accepted as 'scientific fact'.

What does that mean? What is 'scientific fact' and why is it accepted? What is *science* – and is biology like other sciences such as physics (except in subject matter)? The book you are now reading investigates these questions – and some related ones. Like *About Life*, it may particularly interest a reader who wishes to change career to biology and its related subdisciplines. In line with a recommendation by the British Association for the Advancement of Science – that the public should be given fuller information about the nature of science – we present the concepts underpinning biology and a survey of its historical and philosophical basis.

In the first chapter of *About Life* we defined science, provisionally, as *a way of satisfying our curiosity by formulating questions about what we observe and answering them dispassionately, without making value judgements*. That definition seemed adequate at the time, but it is easy to pick holes in it. For example, the word 'science' is used regularly in television programmes, magazines, websites and broadsheet newspapers, but it seems to be used in different senses. How can we interpret the word when its meaning varies?

For most people, most of the time, 'science' means knowledge of a certain sort[1]: a collection of facts and beliefs that helps us to explain and predict the observable world coherently. A science textbook is a repository of such knowledge. When you study science at school or university you learn some of it. But 'scientific knowledge'

---

[1] The Latin *scientia* is usually translated as 'knowledge'. Prior to about 1800, 'science' denoted knowledge and understanding in general; for instance, what are now loosely called 'the humanities' were called 'moral sciences'. In the late 17th and 18th centuries, what we *now* call 'science' was labelled 'natural philosophy'. The word 'scientist' was invented in the 1830s by William Whewell, first president of the British Association for the Advancement of Science, but was not widely accepted until well into the 20th century.

changes continuously. You only have to compare an old edition of a textbook with a recent one to see how much has had to be rewritten in the last 10–15 years.[2] Emerging techniques reveal new facts about the world and our way of thinking has to change to accommodate them. Indeed, many different factors influence the way in which science changes: political, economic, religious, and so on. Therefore, a 'scientific fact' – a 'scientific truth' – is not constant or absolute or 'eternal'. Historians of science can tell us how, and in part *why*, our understanding of nature has changed over time. If we are to understand what science is and in what sense it can claim to provide 'truth', we need to understand why it changes. Therefore, much of this book is about history: the traditions from which modern science evolved and the controversies that arose in the process. Our emphasis from Chapter 6 onwards is on the history of biology.

Practising scientists use the word 'science' to describe their day-to-day work: planning and performing experiments, making observations, recording data, interpreting results, deducing, predicting, speculating, and communicating their findings. Before you are entitled to participate in these activities you must pass a number of examinations and serve what amounts to an apprenticeship under the guidance of one or more established practitioners. You will find yourself facing a career structure with various pay scales and competing, often intensely, with similarly qualified people. A code of professional ethics (largely unspoken) helps to regulate this competition. It should also regulate other aspects of your behaviour; good scientists do not invent data or steal each other's results; and when appropriate, they consider their new ideas in relation to technology and public debate. Understood in this sense – what people called 'scientists' *do* – 'science' is a subject for sociologists.

However, when practising scientists are asked what 'science' is, they seldom answer in terms of their daily work or their ethics. More commonly they tell us that science is a special and distinctive way of thinking about the natural world, unmatched in the intricate detail, practical applicability or 'truth' of what it generates. But what exactly *is* this way of thinking? How is it 'distinctive'? And in what sense is the knowledge it produces 'true'? Most of those are questions for philosophers, but scientists should also consider them.

It is surprisingly difficult to pin down the relationships among the history, sociology and philosophy of science. Sociologists of science look at single frozen frames in the film of history. History illustrates and tests the arguments of philosophers. The history, philosophy and sociology of science are collectively labelled 'science studies', but they remain separate disciplines, each with its own methods and standards of quality. They are specialised subjects, though their domains overlap considerably. In this book, we shall adopt arguments and perspectives from each of them to suit our purposes, but we shall not go into details.

---

[2] A 'scientific fact' may be here today but gone tomorrow as new evidence gives us greater understanding and corrects mistaken notions.

Many of our colleagues, including some eminent ones, have a deep antipathy to 'science studies', which they think distorts our picture of science and its status as a uniquely reliable mode of knowledge.[3] They say that it damages the public image and therefore the funding of science. We understand this antipathy, but the best work in the science studies disciplines should not be dismissed lightly. We need it to answer our questions about what science is, and *why* it is, and to explore the similarities and differences between biology and other sciences. In order to study the thinking behind the science of biology – which includes the whole range of knowledge from early life forms to modern medicine – we have to consider how it arose, and to understand, as well as we can, the thought process and philosophies of the pioneers of modern thought.

Without such considerations, we cannot go on to explore the most controversial topics associated with biology and other sciences today: patenting of human genes, cloning, genetic modification of crops, the obliteration of habitats, the extinction of species, and so on. These are matters that concern everyone, and we all need to be able to discuss them rationally, from an informed standpoint. We offer this book in an effort to meet that need.

The bibliography comprises publications that expand on the ideas presented in the text or offer different perspectives: some are introductory and others are more advanced, but all should be accessible to the non-specialist reader. For particular points, we have relied on professional publications and old or classical works that may interest readers with specialist backgrounds; we have cited these in footnotes at appropriate points in the text rather than adding them to the bibliography.

The book has grown from many years of reading and discussion. Among the numerous colleagues with whom we have exchanged views during the past four decades, Larry Briskman, Jürgen Lawrenz, Christopher Longuet-Higgins, Colm Malone, Jacques Monod, Karl Popper, John Porteous, Lewis Wolpert and J. Z. Young have perhaps been particularly influential. We are also grateful to Lloyd Demetrius and Carolyn Fisher for their helpful criticisms of draft chapters.

May 2008   Paul S. Agutter
                                                                                 Denys N. Wheatley

---

[3] Academic 'disciplines' are artificial categories – labels attached to different parts of a spectrum of human activity. These labels enable us to understand broadly where each individual operates and from what basis their arguments are developed.

Contents

Chapter 1
What is Science?

What we Know

Try making a list all the things you know. You will soon give up – the task would be endless. But even a partial list will show that you have different *kinds* of knowledge.

For example: you know that just now you are sitting down, or standing, or walking across the room. You know the furnishings and the colour of the paintwork. You know what music is playing on the radio. You know, perhaps, that it is a cool day outside but warm in the house, and that you are thirsty. These are immediate, direct *sensations*.

You went to Benidorm on holiday last summer. A friend told you an amusing story yesterday evening. You once fell downstairs when you were four years old. One of your secondary school teachers had a face like a terrapin. You know all these things because you have a *memory*, a gigantic store of past experiences. You remember them.

You know how to ride a bicycle, drive a car, boil an egg, write your name, hammer a nail into a wall. So your list of knowledge includes *skills* as well as immediate and remembered sensations, skills that you have learned and practised.

Two plus two equals four. Dogs bark when they are disturbed or excited. Acorns grow into oaks. Grass is green; water is wet. We turn our experience into *generalisations*, which make up much of our everyday knowledge. Generalisations tell us about patterns and regularities in the world, so we use them to *predict* future events. We also depend on them when we try to devise ways of *controlling* or *manipulating* events and objects.

You know that the Earth is round (more correctly, an 'oblate spheroid'), and that the climate is changing, and that uranium is mined and purified from ores – not because you have experienced or discovered those facts for yourself but because you have been told them by authorities you trust. We gain much of our knowledge by *learning*, usually indirectly, from the people who found it out. Many of our generalisations come from what we hear or read, not from direct experience.

We have moral knowledge (it is wrong to rob banks or set fire to the neighbours' dustbins). We know our family and friends (what they look like, the sound of their voices, how they dress, what work they do, their mannerisms, their attitudes and beliefs). And we have many other sorts of knowledge.

P.S. Agutter, D.N. Wheatley, *Thinking about Life*,
© Springer Science+Business Media B.V. 2008

Few items on this fragmentary list of things-we-know have anything to do with science. So if science is 'knowledge', it must be a special sort of knowledge – but special in what ways?

Scientific Knowledge

At first sight, it is easier to identify ways in which science is *not* special.

Like the rest of what we know, much of it relies on sensory experience, immediate or remembered. We test scientific ideas by matching them against what we can perceive: if the idea fails to match the perception, we reject it. Many scientific 'perceptions' depend on instruments that – as it were – extend the range of what we can sense. Objects too small to see with our unaided eyes can be seen through a microscope. Radiation with wavelengths longer or shorter than light (infrared or ultraviolet, for example) cannot be perceived through our eyes or other senses, but it can be detected and measured by special instruments. 'Measured' is an important word here; whenever we can, we attach numbers to what we (directly or indirectly) perceive. So scientific knowledge includes things that we can perceive using instruments as well as our unaided senses, and – when possible – *measurements* of these things. Fundamentally, however, it depends on what we sense, just like our ordinary everyday knowledge.

It also includes generalisations. Indeed, *specific* statements about individual objects or events have limited value in science. We are mainly concerned with general statements, regular and recurrent patterns in what is perceived. Most of the scientific generalisations we accept are learned from other people's work. That is inevitable; we would never make progress if we had to rebuild scientific knowledge from scratch in each new generation. An important use of generalisations in science – learned or otherwise – is to predict future events; but as we have already said, that is also true of the rest of our knowledge.

Skills are involved, too. A scientist has to learn how to use specialist instruments and understand what they reveal, and must be able to execute experiments. (An experiment is an arrangement by which a particular object or event can be observed and measured without interference from the rest of the world.) But every tradesman makes skilled use of special equipment.

In all these respects, science is only *slightly* different from everyday knowledge. Reliance on perceptions, generalisations and skills hardly makes it 'special'. Science is interesting in the way it is organised (theories, hypotheses and experiments), but even this is not particularly special, as the following everyday analogy shows.

Suppose your car is not stopping as quickly as it should when you apply the brake. You ask why. Perhaps there is insufficient brake fluid; you check to make sure. If your guess was right, you top up the brake fluid. You then ask why the level was low – a leak in the system, perhaps? If your guess was wrong, you try another: maybe the brake linings are worn and need to be replaced. You check (or ask a mechanic to check).

What is happening during this sequence of observations, guesses and checks? First, at the outset, you have an organised understanding of car braking systems and how they work. This 'organised understanding' is what in science we call a *theory*. That may seem an odd use of the word 'theory' since it consists largely of matters of fact, but scientific theories *do* largely consist of matters of fact. Second, you see a problem (the car is not stopping as it should), ask a question about it (why is the car not stopping as it should?), and then use your organised understanding – your theory – to guess a plausible answer. In science, such a guess ('the level of brake fluid is low') is called a *hypothesis*. A hypothesis is a possible answer to a question; it must be consistent with your theory and – crucially – it must be *testable* (in this case, by checking the level of brake fluid). If the test proves your guess wrong, i.e. the hypothesis is *refuted*, you try another guess – an alternative hypothesis. If the test seems to confirm your guess, you ask a further question (why was the level of brake fluid low?), propose a further hypothesis consistent with the theory (a leak in the system), test it (by looking for the leak), and so on.

Although this failing-brakes analogy is much simpler than most scientific reasoning, it is the same in principle. A scientific investigation entails exactly the same steps: noticing something odd or interesting, asking questions about it, proposing hypotheses (i.e. guessing possible answers) in the context of the relevant theory, and testing the hypotheses. That is how science progresses.

Notice the *nature* of the hypotheses in our analogy. They link a possible *cause* (low brake fluid, worn brake linings) to an *effect* (impaired brake function). All scientific hypotheses have that character: they are provisional cause-effect relationships that seem plausible in the light of an accepted theory (organised existing knowledge).

Cause-effect reasoning is the foundation of our understanding of the world, not least our scientific understanding. It enables us to make predictions. It enables us to find ways of controlling and manipulating events and objects.

The Need to Understand Cause and Effect is Uniquely and Characteristically Human

Chimpanzees seem to have only a limited idea of cause and effect. They can use simple tools such as sticks to recover food that is out of reach, but given a selection of possible tools, they cannot decide which is best for the job. They can pile boxes on top of one another to obtain a food reward, but they never realise that if the floor is uneven the boxes will topple over. A young human child might make such a mistake but will quickly learn from it. Chimpanzees can master human language to some extent, but they do not seem to use language in the wild. Human children acquire language in the first few years of life. Language is crucial for our understanding, not least our ability to think in terms of cause and effect and to learn skills.

Humans are very closely related to chimpanzees but our mental capacities are qualitatively different. We can infer cause-effect relationships, predict events, and express them in language. Our capacity to acquire skills, too, is much superior to that of other species. We touched on the question of human uniqueness in *About Life* – it may be rooted in our social nature and our bipedalism – but whatever the reasons, our mental and technological capacities are indisputably unique. We depend on those capacities throughout our lives and we prize them highly. They make us human.

Equipped with such capacities, our early ancestors must have found themselves in a world full of objects and events for which there were no evident causes: changes in the weather, the cycle of seasons, volcanoes, birth and death. When we cannot understand the world around us in terms of cause and effect, we do not know how to solve day-to-day problems and we cannot decide what actions to take. We find such uncertainty intolerable. So our ancestors *invented* explanations for mysterious events; they constructed theories. Those theories probably took the form of myths (explanatory stories). We cope poorly in the absence of some form or system of beliefs – we need to tell stories that purport to explain what happens around us and to us. The price of our uniqueness seems to be a *need* as well as an *ability* to explain causes and effects.

What sort of beliefs might our ancestors have formed? The only causal agents they knew from direct experience were animate. Humans and other animals could deliberately cause things to happen, but inanimate objects could not. So all unexplained events may have been attributed to animate, perhaps human-like and almost invariably invisible agencies, which might be appeased or rendered co-operative by appropriate rituals or sacrifices. Early human groups who adopted such a policy probably fared better than those who did not: it would have made for greater social cohesion and enabled them to cope practically with the uncertainties of an unpredictable and often hostile world. It would at least have reassured them. Belief of this primitive religious sort could therefore have become selectively advantageous for early humans.

If so, then we, their descendants, may have inherited a biological predisposition to tell ourselves stories that make sense of our lives and the world around us. Faced with serious uncertainties, mortal dangers or traumas, we tend to become more religious. People with strong religious beliefs often lead longer and healthier lives and deal more effectively with adversity than sceptics do. Belief seems to be *biologically necessary* for humans, and 'supernaturalistic' or religious belief seems to be the most biologically fundamental (and therefore much the most widespread) kind.

'Belief' in this context encompasses all the contents of our minds and memories: what we know and understand, how we explain, predict, inquire, and so on. It also includes the articles of faith on which our story-telling depends. Modern science, too, is founded on articles of faith, namely that the universe we observe is intrinsically orderly and that human minds are capable of grasping the essence of that order.[1] These are plausible *but untestable* premises.

---

[1] Einstein remarked that '*The eternal mystery of the world is its comprehensibility*'. But these two articles of faith have not been evident in all cultures.

What is Distinctive About Science?

The function of science is to make sense of the world so that we can answer questions, make predictions and control and manipulate events. But every human culture has its own ways of making sense of the world and of controlling and manipulating it. So what *is* different about science?

So far, we have found that scientific knowledge is broadly similar to everyday knowledge. Reading between the lines, however, we can begin to see distinctions.

Science is Naturalistic Not Supernaturalistic

Supernaturalistic belief, which we have argued is the most basic and widespread kind, differs from science. A supernaturalistic system seeks human, demonic or theistic causes for all events – diseases, for example. As scientists, we attribute diseases to infectious organisms, gene defects, environmental toxins and so on. The orbit of a planet and the progress of an avalanche are explained in terms of antecedent causes and theories of mechanics, not the dispositions and whims of gods. Science presupposes that *causes lie in the observable material world itself.* That distinguishes a 'scientific culture' from the overwhelming majority of human cultures.

Attempts to find naturalistic explanations date back to Classical Greece, which is why histories of western science traditionally begin with a glance at the Greek and Alexandrian philosophers. It is interesting to ask *why* naturalistic belief began in Greece two and a half millennia ago, and *why* the modern developed world has adopted it.

Scientific Explanations Are Mechanistic

Science explains events in terms of mechanisms. The cause must be a physical situation existing *before* the event, and it must have nothing to do with motives or intentions or purposes. If events are caused by conscious, animate, human-like agencies, then to explain the causes we must understand the motives – the purposes – of those agencies. If the causes are mechanistic, then it is misleading to consider motives and purposes because there *are* no motives and purposes. Events in the natural world have no intentions behind them.

You might suppose that if an explanation is naturalistic then it must be mechanistic as well. However, the world-view of the great Greek philosopher Aristotle was entirely naturalistic, but his explanations were deeply *teleological*, i.e. expressed in terms of purposes. As we shall see, the emergence of modern science, especially biology, entailed a long love-hate relationship with Aristotle's teachings.

It might seem impossible to eliminate teleology from biology because biological entities – parts of organisms – *do* have purposes. Nevertheless, modern biology is a science; its explanations are as mechanistic as any in physics. This seeming paradox, and its resolution, will dominate much of this book after Chapter 7.

Scientific Ideas Are Expressed in 'Value-Neutral' Terms

In science, we do not evaluate things as good or bad, desirable or undesirable, beautiful or ugly, etc. We say what *is* the case and how it fits or fails to fit our existing theories. At least, that is the accepted ideal. Privately, we may (and often do) consider the result of an experiment good or bad depending on whether it is consistent or inconsistent with a favourite hypothesis. The hypothesis may have become a favourite because we find (or found) it aesthetically pleasing.

The claim that science is value-neutral is contentious; sociologists of science find compelling reasons to dismiss it. For example, what we actually choose to study, and the way in which we study it, may be determined largely by political, economic and other influences outside science. But any *particular* piece of scientific discourse is value-neutral, or should be.[2]

Scientific Explanations Are General Rather than Particular

Explanations in most cultures have tended to be particular rather than general. For example, the question was not 'What causes boils?' but 'Why has *this* person, or *this* group of people, been afflicted with boils at *this* particular time?' The answer was expressed in terms of particular practices and moral codes, or particular enmities. The cure would lie in countering a witch's power or appeasing an appropriate deity or practising some sort of incantation. Scientifically, we explain *all* cases of boils in terms of a common pathogenic agent. The cure lies in an antibiotic suitable for treating all cases of *Staphylococcus aureus* infection. The explanation is *general*.

---

[2] To see how contentious this matter can be, consider the following dilemma. All reasonable people consider racism to be morally unacceptable. However, many biologists (including psychologists) believe that human behaviour is mostly determined by our genes. That includes criminal behaviour. Now, it is well known that a disproportionate number of the inmates of American prisons are black. So it seems scientifically reasonable to infer that black Americans are genetically more predisposed to crime than white Americans. How do we handle this inference? We can decide either that the premise is wrong – i.e. that criminality is *not* significantly influenced by our genes – or that science compels us to be racist. If we choose the former alternative, we risk being accused of allowing values to influence our scientific judgment. (The authors of this book, incidentally, are willing to take that risk!)

Scientific Explanations Tend towards 'Reductionism'

In science, we usually try to explain large-scale phenomena in terms of smaller-scale parts. For instance, we seek to understand living bodies in terms of the cells they comprise, each cell in terms of its component molecules, and each molecule in terms of its atoms. The 'reductionist' approach helps us to explain things mechanistically and in the most general terms. Everything in the world is made up of the same small selection of atoms, in various combinations, so at the atomic scale everything must obey the same laws. The smaller the scale we observe, the more wide-ranging the phenomena. It is *large*-scale entities that are individual and particular.

Scientific Explanations Seek to Be Comprehensive

This 'reductionist' tendency helps us to find the most widely-applicable patterns of understanding (i.e. theories). Scientific theories aim at *comprehensiveness*, unifying our knowledge. We look for similarities underlying apparently *disparate* phenomena such as the falling of an apple and the motions of the planets – which would make no sense in most cultures. This is a hallmark of science. Like naturalism, a quest for comprehensive theories is evident in the work of the Greek philosophers, most obviously Aristotle. In contrast, other cultures tend not to seek general, widely-applicable theories; they are more concerned with the individual and particular and readily accept the *ad hoc*.

Scientific Explanations Tend to Be Abstract and, where Possible, Mathematical

Particular descriptions of objects and events are concrete and (usually) qualitative. We can only connect the falling of an apple with the motions of the planets when we focus on the *abstract* similarities between these events. Mathematics is a very effective way of expressing such abstractions precisely and deducing predictions from them. If we measure things rather than simply describing them – if we find out how *quickly* the apple falls – we can test those predictions critically.

Science Aims to Be – and to Make Things – as Simple as Possible

This claim may seem surprising, but it is true – provided you do not confuse 'simple' with 'concrete'. In science, redundant ideas are eliminated. The causes proposed for any phenomenon are pared down to the bare minimum. *Parsimony* is a hallmark of a good theory, and of a good hypothesis: the simpler the suggested cause, the easier it is to test it unequivocally.

Scientific Theories Must Be Logically Consistent

Logical inconsistencies (self-contradictions) are eliminated by the abstraction and paring-down processes, but scientific explanations must also be *plausible* and *cogent*. They must be consistent with the rest of our beliefs – particularly with other scientific theories.

In Principle, Scientific Knowledge Is 'Public'

The means by which scientific explanations are established and tested must be *open to public scrutiny*, at least in principle. Divine inspiration and individual imagination are not acceptable criteria for belief, as they have been in many cultures. No matter how a scientific idea originates, which might indeed be a matter of individual imagination[3] (hunches), the observations, experiments and reasoning involved in testing it must be reproducible by anyone (provided he or she is appropriately qualified).

Scientific Knowledge Is Impersonal

For this reason, scientific writing is never couched in personal terms; it seeks to be *objective*. As scientists we write 'A causes B', 'Event X happens in such-and-such circumstances', 'Objects such as Y have such-and-such properties', and so on. We do not use 'I believe…' or 'I observed…', even though what we write about is the product of individual human senses and human thought. That is why scientific English is so peculiar: passive voice, few adjectives and adverbs, no rhetorical colour. It is dry and 'factual' – it would be difficult to read, largely devoid of charm, even without the formidable technical vocabulary. Individual personalities, and personal friendships and enmities, should not be allowed to influence science – though they sometimes do, e.g. in 'personal' remarks in scientific reports on research activities that are trying to establish new facts.

---

[3] Or a dream. Famously, the great organic chemist Kekulé solved the structure of benzene by means of a dream – but the 'dream-structure' then withstood critical experimental testing. It was the critical testing, not the dream, that made Kekulé's solution scientific and entitles us to believe it.

Scientific Knowledge Is Inherently Progressive

A scientific hypothesis must be tested, and when you test it you may make novel observations and discover new phenomena through your research work. The knowledge and beliefs that arise during the search for scientific explanations change over time. Most cultures undergo changes in their belief systems, largely as a result of contact with other cultures, but scientific knowledge is *inherently progressive and permanently provisional*. It is never intended to be as the Laws of the Medes and Persians – though it is often treated as though it were. What we know and believe now is not what we knew and believed in the past – and we will know and believe something different again in the future.

Why Is Science Distinctive?

In many general ways, science is just like everyday knowledge: it depends on sensory experience, memory, generalisations, skills and the search for cause-effect relationships. In other respects, it is unlike the systems of knowledge and belief in other cultures: it is naturalistic, mechanistic, value-neutral, general, reductionist, comprehensive, abstract, simple, logically coherent, 'public', impersonal and inherently progressive.

These distinctions are crucial. The findings of science seldom match intuition. Everyday knowledge does not tell us that the Earth orbits the Sun or that humans share half their genes with bananas, but science persuades us. Electrons, quasars and hormone receptors are not objects familiar from everyday life but they are familiar elements of scientific discourse. A way of producing knowledge that generates data so remote from sensory experience, and beliefs so contrary to intuition, is distinctly peculiar.

If science is such a peculiar sort of knowledge, so different from knowledge in other cultures, we would expect it to have arisen very infrequently during the course of human history, and only in particular locations.

That is a definite *prediction*. We can test it by examining the evidence.

Science Originated and Flourished in a Particular Time and Place

To find that evidence, look at the major topics covered in science textbooks (school or undergraduate level). For each topic, note the country and the century in which the basic ideas were established. Table 1.1 is an example. When the items in Table 1.1 are displayed on a map of the world (Fig. 1.1) and on a time-scale of human history (Fig. 1.2), the result corroborates our prediction – rather strikingly.

Table 1.1 Origins of some general topics in science

Topic	Country(ies) of origin	Century of origin
Human anatomy	Italy	16th
Heliocentric solar system	Poland	16th
	Italy, Czech Republic	17th
Geometrical optics	Holland, England	17th
Classical mechanics	Italy, France, England	17th
Biological microscopy	Holland, England, Italy	17th
Electrostatics	Italy	18th
Biological taxonomy	Sweden	18th
Electrodynamics	France, England	19th
Thermodynamics	England, Austria	19th
Electromagnetism	England, Scotland	19th
Atomic theory	England, Sweden	19th
Physical chemistry	Germany	19th
Organic chemistry	Germany, France	19th
Periodic table of elements	Russia	19th
Bacteriology	France, Germany	19th
Evolutionary theory	France, England	19th
Relativity theory	Switzerland	20th
Quantum mechanics	Germany	20th
'Big bang' theory of cosmology	USA	20th

Fig. 1.1 Outline map of the world showing the places of origin of the topics listed in Table 1.1

The inference is irresistible: all the topics that we regard as parts of 'science' had their roots in Europe (predominantly northern and western Europe) between the 16th and the 20th centuries – and apparently in no other place and at no other time.

You may wish to examine a different sample of topics. The result will be much the same, though if you choose disciplines of very recent origin then you might find

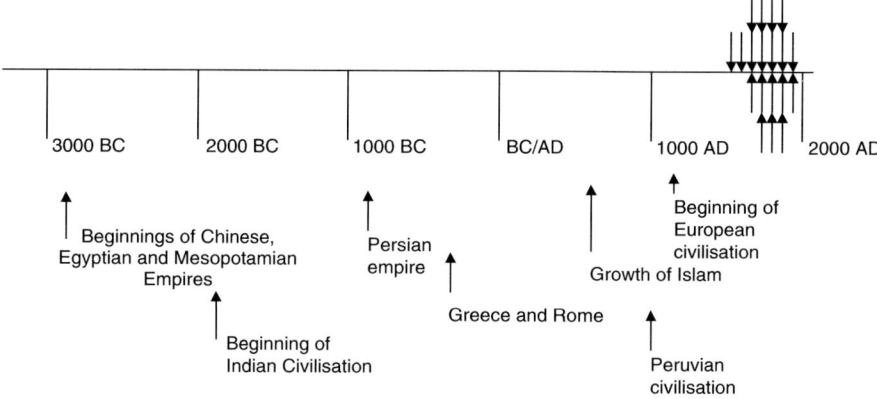

Fig. 1.2 Time-line of the history of human civilisation. The approximate dates of origin of major civilisations are shown in the lower part of the illustration. The arrows clustered above and below the extreme right of the time-line show the approximate dates of origin of the topics listed in Table 1.1

more of them appearing in North America or some other part of the developed world culturally rooted in European colonisation. None of the sets of ideas we now recognise as 'scientific' came from the great civilisations of ancient Egypt or Mesopotamia or India or China – though a great deal of knowledge from those civilisations came down to us via trade links. None of our modern scientific disciplines arose in Classical Greece or Rome, despite the naturalism of Greek thought and despite the great shaping influence that the Classical world exerted on later Islamic and European civilisation. There was nothing quite akin to modern science in Europe itself during the period 1100–1500, though European thinkers at that time had important insights into what we now call 'science'.

How can we account for this? Why did so curious and 'unnatural' a way of seeking knowledge as modern science arise in 16th and 17th century Europe, but apparently in no other place and at no other time in history? And why do we consider it more reliable than the more widespread (supernaturalistic,[4] specific, concrete) alternatives?

While we explore these questions, we should ask whether our data and reasoning are sound. Could appearances have deceived us? Was there, after all, a precedent for what we now call science?

---

[4] The word 'supernatural' is not used in modern science, though it remains extant in the vernacular. A scientist would qualify any mention of it by saying that it is a term used to describe a phenomenon that has yet to be explained rationally. One possible example is the alleged 'telepathy' between identical twins, the apparent ability to 'communicate' mentally over thousands of miles or behave simultaneously in very similar ways.

Chapter 2
Culture, Technology and Knowledge

We face three interrelated questions. First, is science uniquely the product of post-mediaeval Western Europe? Second, if not, where (and when) else did it arise – and why do we never hear about it? Third, if the naturalistic learning of Classical Greece was not science, how did it come to exert such a profound influence on the *emergence* of science in Europe 1,800 years or more afterwards? To tackle these questions, we first need to examine: (1) the relationship between knowledge or belief and culture, and (2) the routes by which Classical learning reached late mediaeval Europe.

Ways of knowing in all cultures share common features (sensory experience, memory, generalisation, mastery of skills, search for cause-effect relationships, etc.) because they are all human; but each is also distinctive. By definition, different cultures arise in different environments, each with its particular history. Tropical desert dwellers are unlikely to have knowledge and beliefs about polar bears or snow, but they are likely to know a good deal about sand, dehydration and scorpions. In the modern developed world, a hill farmer has different knowledge and beliefs from, say, a city accountant.

In other words, while humans are biologically predisposed to gain an understanding of the observable world, the *particular* beliefs they form and the *particular* knowledge they acquire depend on where and how they live. Can we say anything about the relationship between life-style and knowledge that will help us to understand why science arose when and where it did – what it was about post-mediaeval Western Europe that caused, or enabled, scientific thought to emerge?

Archaeologists and social anthropologists understand a lot about the differences among cultures, but they are reluctant to generalise. However, relatively little of their work concerns knowledge or beliefs (which in any case are not accessible to archaeologists, except where they are investigating a culture that left written records). Their focus is rather on technology: tools, weapons, what people wore and what they ate and what their homes were like. Appropriate tools, clothing, housing and methods for obtaining and preparing food are necessary for human survival, but what is 'appropriate' depends on the environment. Like knowledge and belief, it is culture-specific. When children are brought up, they learn about the technological artefacts of their community as well as information about the world around them. Children in our society, for example, acquire knowledge and beliefs about

P.S. Agutter, D.N. Wheatley, *Thinking about Life*,
© Springer Science+Business Media B.V. 2008

televisions and computers as well as about the structure of the solar system and what genes are.

That raises another pertinent question: what is the relationship between science and technology – or, more generally, between knowledge or belief and technology? It is commonly assumed that, in our modern world, science gives rise to technology. In other words, the idea comes first and the application follows. In fact, however, knowledge and technology can be interconnected in quite complicated ways – or entirely unconnected.

Evolution of Beliefs in Civilised Cultures: A Hypothetical Scenario

The earliest civilisations[1] arose on the fertile flood-plains of great rivers, where sufficient food could be produced every year to support large, non-migratory populations. We know too little about the early history of any ancient culture to describe it in detail, so the following scenario follows no actual historical sequence. However, the pattern of development seems generally correct; the civilisations of ancient China, India, Mesopotamia and Egypt all had broadly similar characteristics. The same applies to Peru, where the immediate environs of the short but rich rivers were home to the earliest known indigenous civilisations of the Americas.

If a civilisation was to be viable, enough food had to be produced to support the large, concentrated populations of one or more cities. Therefore, appropriate times for preparing the land and sowing crops had to be decided. To make the right decisions, the annual flooding of the river had to be predicted. This required a calendar, which was necessarily based on observations of the stars, since the annual cycle of the heavens was the only reliable way of *measuring* – that all-important requirement – the passage of time. The existence of civilisation therefore presupposes an astronomically-based calendar.

The fact that such calendars worked reliably must have suggested a *causal* connection between particular configurations of stars and particular terrestrial events, especially the flooding of the river. How could such a connection be explained? The answer would have been couched in terms of deities associated with stars, planets, rivers, crops and so forth. But that would have led to another question: how could the *continuation* of this causal link – vital for the survival of the community – be ensured? Thus, the dawn of civilisation was inexorably linked to particular sets of knowledge, beliefs and practices. These included the names and natures of gods and the rituals appropriate for their worship and appeasement, as well as the study and interpretation of the night sky and the construction of calendars.

---

[1] Here we intend 'civilisation' in its literal sense, i.e. a community living in or centred on one or more cities.

No less important was social organisation. To ensure that sowing and harvesting were timed properly and conducted efficiently, a social hierarchy was needed. Civilised society needed leaders, just as it needed experts to interpret the heavens and mediate with gods. It certainly needed many labourers and someone to organise them. Kings, priests (kings usually being representatives of a priestly class) and numerous agricultural workers were inevitable concomitants of early civilisation.

Equally inevitable were improved tools for planting and harvesting and, in due course, for food warehousing and distribution. The better the tools, the bigger the harvest and the faster the population growth; and the bigger the harvest and the population, the greater the need for organised distribution. The dawn of civilisation had inescapable implications for technology and social organisation as well as for knowledge and beliefs.

It was natural[2] to generalise the causal link between celestial and terrestrial events: perhaps *all* significant terrestrial events were heralded by celestial ones. In particular, rare and dramatic celestial events such as comets and eclipses betokened rare and dramatic terrestrial occurrences such as the fall or birth of kingdoms. If so, then the more knowledge the priests gained about the stars, the more they and the king would be able to control – or at least anticipate and perhaps forestall, i.e. control – events on Earth. They pursued more detailed studies of the apparent movements of stars, planets, sun and moon. Astronomy, or more particularly astrology, was an early development in all ancient civilisations.

These studies entailed mathematics. Mathematical techniques were needed to describe the patterns of movement of celestial bodies and to use them as the basis for personal and political predictions. Mathematics also proved invaluable for civil engineering projects such as pyramid-building and for calculating food rations in times of shortage.

The (usually) rich food supply of a civilised community attracted the attentions of nomadic and other pilferers from outside as well as from greedy and nefarious persons within. It must soon have become necessary to formulate laws relating to defence of the community and its food supplies. Such laws were no doubt authorised by gods. It was also necessary to construct systems of defence including city walls, strong buildings, weapons, and at least a potential army to deploy them. Lawgiving and defence[3] became increasingly important facets of the work of kings.

---

[2] We all have a 'natural' – a biological – tendency to seek causal connections and to generalise. If event B follows event A, even if there is no obvious connection between them, we are inclined to believe that A causes B, especially if the coincidence happens more than once. This 'default' type of thinking can sometimes give us useful knowledge, but it often misleads or generates superstitions. If you walk under a ladder and the person atop the ladder drops something on you, you are likely to infer that walking under ladders causes bad luck. As scientists, we learn to be cautious about such connections.

[3] 'Defence', of course, is a well-known euphemism. A community with sophisticated military organisation and weaponry has an irresistible urge to conquer and subjugate its less fortunately equipped neighbours. The need for defence merges very quickly into aggressive expansionist policy, and the military hero is the focus of most ancient-world epics. *Plus ça change*. It is a fact that most of the longer-lived ancient-world civilisations expanded by conquest.

Early civilisations therefore gave rise to innovations in building and military as well as agricultural technology, and to systems of law and polytheistic religions.

As the astrological and mathematical knowledge of the priests became more extensive and complicated, the system of law continued to grow, and the need to record the production and distribution of food and weapons increased, writing became necessary. All ancient civilisations produced some form of writing.

Dense, city-dwelling, non-migratory human populations invariably develop epidemic diseases. Medicine became essential to prevent population collapse and some individuals began to specialise in its practice. With the development of writing, medical treatments for various diseases were recorded for posterity.

To summarise (Fig. 2.1): early civilisations produced polytheistic religions, calendars, astronomy, astrology, mathematics, medicine, law, writing, social stratification, division of labour, civil engineering and agricultural and military technology. But there was no direct connection in our hypothetical scenario between the knowledge and beliefs of the community and its technologies. They were the provinces of different social groups. Knowledge and technology are both intimately related

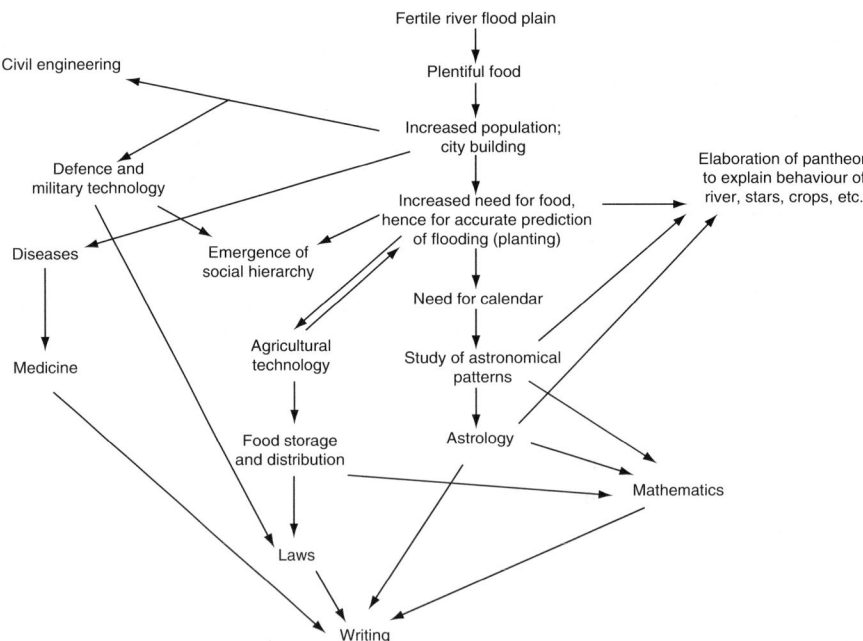

Fig. 2.1 Evolution of a hypothetical ancient civilisation. The flow diagram illustrates the scenario outlined in the text. Broadly, 'technology' and 'social organisation' are on the left of the picture and 'knowledge' is on the right. 'Medicine' was a practical art or craft and is therefore better classed as technology than as knowledge. There is relatively little interaction between the two sides of the diagram, with one important exception (not shown, since the picture is already complicated): the gods in the pantheon would be evoked to legitimise the social hierarchy and any civil engineering and military developments, as well as to maintain social cohesion

to cultural evolution and its socio-economic basis, but the association between them is often indirect and elusive.

Modern Science and Technology

How mutually dependent are knowledge and technology in the modern world? As we hinted earlier, the belief that *science begets technology* (the idea always precedes the application) is dubious. Of course, it is the standard justification for both public and private funding of scientific research: finance science and it will produce new ideas, some of which will be applied in new technology, which will generate profit and boost the economy. If the premise is dubious, so is this justification for funding research. Let us consider a few examples.

Optical Instruments

The compound microscope and the telescope were first made in the Netherlands in the late 16th century, probably as showpieces by the master lens grinders who flourished in Flemish towns.[4] The theory of geometrical optics, which explains how these instruments work, was constructed 20 to 30 years *later*, again in the Netherlands. Galileo improved the telescope and obtained revolutionary new data about the solar system. Later in the 17th century, microscopists (many of them Dutch) used the microscope to obtain dramatic and wholly unexpected insights into the living world.

These examples show that new technology does not (always or usually) come from science, but rather from pre-existing technology – in this case, the craft of lens-grinding. Moreover, the new optical technology of the 16th–17th centuries gave rise to new knowledge in two distinct ways. First, the telescope and microscope produced observations about the solar system and about the structure of living matter that could not otherwise have been obtained. Lens grinders, not scientists, had provided us with the means to 'think big' and 'think small'; their instruments had enabled us to *see* big and small for the first time. Second, a new theory – geometrical

---

[4] The Flemish merchants were Protestant, and Protestants had an obligation to read the Bible in their own language. Printed copies of the Bible were widely available. But the merchants' houses had large rooms that were quite dark for much of the year, so people with imperfect eyesight struggled to perform the basic religious duty of reading the Scriptures. This created a market for spectacles, which fostered the profession of lens-grinding. Master lens-grinders competed with each other and the competition enhanced their skills. Of course, the historical process was more complicated than that, but it is no coincidence that lens-grinders achieved such near-perfection in late 16th century Flanders. Lenses to aid vision had first been developed in Islamic Spain by Ibn Firnas (810–887) and were later described by Roger Bacon (c. 1214–1292).

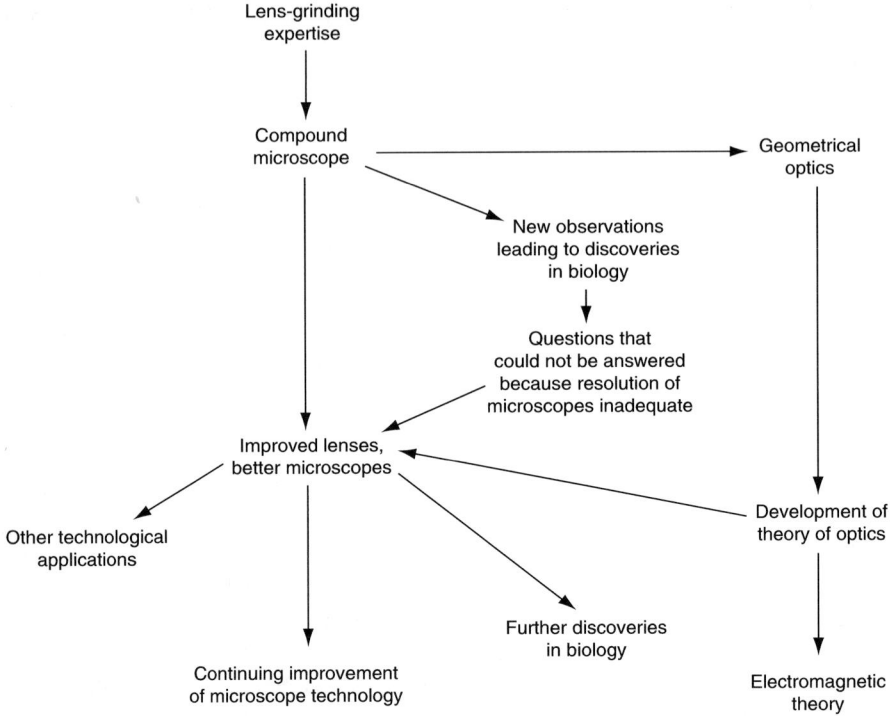

Fig. 2.2 Co-development of technology and knowledge. The example of the compound micro-scope (we could equally well have used the telescope for illustration) shows, in simplified form, the type of 'dialogue' between technology and knowledge that typifies the progress of science. Again, technological developments are indicated on the left of the picture and advances in knowl-edge on the right. Note that these advances include both biology (observations under the micro-scope) and physics (theories of optics and electromagnetism). In fact, optical microscopes improved in stages; salient advances included the introduction of the achromatic lens in 1827 and the Abbe lens (eliminating spherical aberration) in the late 1870s

optics – was elaborated to explain how these new instruments worked, and it proved to have wide applications. Thus, technology had revealed new information *and* given rise to a new theory. (In detail, the process was more intricate; see Fig. 2.2.)

In this case – and there are many similar ones – technology begat science, not the other way round.

The Boulton-Watt Steam Engine and the Discovery of Latent Heat

The separate condenser that revolutionised the design of the 18th century steam engine was patented by James Watt. At about the same time, Joseph Black pub-lished his investigations into latent heat. The concept of latent heat explains the operation of the separate condenser. Black and Watt were friends, both citizens of Edinburgh during the Scottish Enlightenment, and they doubtless discussed their technological and philosophical interests on many occasions. The idea of latent heat and the design of the separate condenser probably evolved together as Black and Watt drew inspiration from each other. Typically, scientists and engineers alike kick new ideas around, gradually altering and elaborating them until they become plausible and useful. In the Black-Watt case it is unlikely that science begat technology, or vice versa; the science and the technology probably co-evolved. Such co-evolution, or synergy, typifies progress in any discipline and continues indefinitely.

Electrical Power and the 'Second Industrial Revolution'

Michael Faraday produced a mathematical theory that unified the studies of magnetism and electricity, which was a work of genius; but he also saw practical potential in his electrical generators and motors. Electrical machinery provided a more efficient alternative to steam engines: the 'second industrial revolution'. Here, in a sense, science *did* beget technology. However, Faraday's innovations were incorporated into industry only because there was a factory-based industry in the first place. He could scarcely have envisaged the practical applications of his devices had the first industrial revolution not already happened. New knowledge can be exploited more easily when the required level of technology pre-exists.

The Electron Microscope

An electron microscope works by focusing a beam of electrons, just as a light microscope works by focusing a beam of light. To understand the principle of electron microscopy, you have to understand that electrons can behave as waves as well as particles – a basic idea of quantum mechanics. But the inventors of the electron microscope, Ruska and Knoll, had apparently never heard of quantum mechanics until after they had created the instrument. They were both employees of Siemens in the 1930s, one working on improvements in cathode ray tube design, the other on the elimination of fast electrical transients from power cables. In the course of their work they discovered that a beam of electrons sometimes generated a highly magnified image of the carbon anode on a cathode ray screen. This observation was pure serendipity, but they developed it, and Siemens patented the world's first electron microscope in 1937. Only afterwards did Ruska and Knoll hear about quantum mechanics.

A revolutionary new theory – which had been developed in complete isolation from Siemens or any other firm – was uniquely capable of explaining how a revolutionary new scientific instrument worked. There was no known exchange of information or personnel between the science and the technology, though both quantum mechanics and the electron microscope were largely or exclusively German innovations. This example of a 'coincidental' science-technology relationship is not unique. Comparable circumstances attended the creation of the Newcomen-Savery pump in the early 18th century.[5] Such coincidences merit study. They have not been satisfactorily explained.

The 'Science-Begets-Technology' Myth

The myth that science leads to technology was explicitly articulated by Francis Bacon early in the 17th century (see Chapter 6). He asserted that the quest for knowledge of the natural world would give mankind greater power and control over that world, to increasing human benefit. In other words, the production of new technology was the *purpose* of what we now call science. Scientists, politicians and others have continued to promulgate this Baconian myth ever since and it has seldom been challenged. But as the above examples show, it is far from reality. Technology and science are very closely allied in our society, but the relationship is not one of simple cause and effect. It is complex, interactive, variable and in some cases resistant to analysis.

Yet this close connection between knowledge and technology is notable. As we have seen, technology and knowledge developed more or less independently in ancient civilisations, even when they were rooted in the same culture. What made our culture different? We shall have to defer that question until later.[6]

Origins of Naturalistic Thought

Let us now return to more ancient history. Around 500 BC,[7] a novel way of thinking began to emerge among the Greek-speaking peoples of the Mediterranean world. For the first time in human history (as far as we know), attempts were made to

---

[5] Although Papin's work on pressure vessels provided a theoretical explanation for the way in which this pioneering steam engine worked, it is unlikely that Newcomen or even Savery knew about Papin. Newcomen used a round cylinder because brewers stored beer in round barrels, so he knew the design was good for withstanding pressure: technology was borrowed to develop further technology, and science had little or nothing to do with it.

[6] Our main concern in this book is to examine the emergence of *biology* as a science and to consider the similarities and differences between biology and the physical sciences. It is perhaps worth observing that the 'technologies' traditionally associated with 'biology' (medicine, agriculture, traditional industries such as baking and brewing) are much more conservative in character than the new manufacturing industries that became associated with physics and chemistry at the time of the industrial revolution.

[7] It is intriguing, though probably coincidental, that belief systems throughout the world seem to have undergone a sudden transition towards greater abstraction around 600–500 BC. Naturalistic

explain the existence and character of the world in general, abstract terms rather than by explicit reference to deities or spirits. We know tantalisingly little about these first essays in naturalistic thought. The works of the so-called 'pre-Socratic philosophers' such as Thales, Empedocles, Democritus and Pythagoras are available to us only through isolated fragments or accounts written many generations later. What we *can* say for certain is that the tradition they initiated culminated in the philosophical masterpieces of Classical Athens, the writings of Plato and Aristotle, and that these were to become the fountainhead of modern western thought.

We have argued that in all ancient civilisations, observable phenomena were explained in supernaturalistic terms, by reference to a polytheistic religion. Indeed, the tendency to construct our knowledge and beliefs about the world in supernaturalistic terms seems common to almost all cultures. So why did *naturalistic* thought arise among the Greek-speaking islands and city states of the eastern Mediterranean around 500 BC?

There is no definite answer, but we can speculate on the basis of what we know about those Greek states. How did they differ from ancient civilisations of the kind we envisaged at the start of this chapter? A few points spring to mind:

- Power and wealth in the Greek world seems to have belonged as much to merchants as to priests, war-leaders and land-owners (there was no priestly caste). The valuation of goods for trade requires an ability to relate disparate items mathematically. This may have fostered the development of abstract thought.
- The Greek communities had trading and cultural contact with *several* great empires (Egypt, Babylonia, Persia) that differed in their religions but had similar or complementary knowledge of astronomy and mathematics.[8] They became cultural melting-pots in which astronomical, mathematical and other aspects of knowledge from many cultures could be assimilated, but divorced from its original supernaturalistic contexts.
- For geographical reasons, they did not become the nuclei of expanding military empires. They remained small, decentralised and non-bureaucratic. These conditions favour freedom and independence of thought.
- They competed with each other, which fostered ingenuity.
- They had an alphabetical system of writing, a fairly recent innovation; the earliest alphabet in the world dates from late in the second millenium BC (writing had begun in Mesopotamia and China before 3000 BC). Alphabets help to produce words for abstract ideas; picture-writing tends to limit discourse to the concrete.

---

thought appears to have been unique to the Greek world, but this is also the time of Confucius and Lao Tze in China, Gautama (the Buddha) and the anonymous author of the Upanishads in India, Zoroaster in Persia, and the authors of 'Second Isaiah' and other works of mature Judaism in the Middle East. This pan-cultural transition of human consciousness is well known (Karl Jaspers may have been the first to draw attention to it), but no one knows how to explain it except in terms of coincidence.

[8] For example, the famous 'theorem of Pythagoras' was known to the Babylonians many centuries before the time of Pythagoras, and at least some special applications of it were known to the Egyptians. There seems little doubt that Pythagoras learned of the theorem from these ancient sources and found a new general proof for it.

Some such combination of factors may have facilitated the development of naturalistic thought. On the other hand, almost all these factors had applied in the Phoenician cities Tyre and Sidon several centuries earlier, and perhaps in their colonies such as Carthage in the western Mediterranean. (Perhaps there was not so much competition among the different Phoenician communities, but they had merchant power, wide-ranging cultural contacts and the world's first alphabet, and they did not have military empires to compare with Egypt or Babylon.) Maybe naturalistic thought did begin in Phoenicia, but there are no written records or other evidence to tell us so. Maybe the Phoenicians influenced the development of Greek naturalistic thought, but again we have no evidence. So we cannot be sure whether our 'set of conditions predisposing to naturalistic thought' is either necessary or sufficient. At least it is a plausible guess.

Greek Naturalistic Thought Was Not Science

As far as we can tell from the surviving fragments of their work, the pre-Socratic philosophers developed systems of explanation for the observable world that were not only naturalistic but (partly because of their naturalism) abstract, comprehensive, consistent, parsimonious and to an extent reductionist. (The most strikingly reductionist example is the philosophy of Democritus, who pronounced that all matter consists of ultimate indivisible units, *atoms*.) However, they were teleological rather than mechanistic; i.e. they admitted purpose and meaning in natural phenomena. They were open to critical scrutiny only by logical debate, not by experiment. They were not inherently progressive in the sense that modern science is – by advancing, testing and replacing hypotheses – so they remained 'fixed' or static. Nor were they fruitful in terms of prediction or application. In short, although they shared some of the distinctive characteristics of modern science (Chapter 1), they by no means shared all of them. To call ancient Greek writings 'science' is therefore a misnomer.

We recognise ancient Greek culture as the cradle of poetry and a rich source of myths, so how central to it was 'naturalistic thought'? Greek poetry and legends from the time of Homer, the 'Teacher of all Hellas', concerned the forms, limits, laws, rationale and order of the cosmos. Poetry was an integral part of a gentleman's education because it conferred understanding of these matters. Hesiod's genealogical account of the gods was designed to explain how order was imposed on the world. The early Greek philosophers saw that forms, limits,[9] order, law, reason and order could be explained without reference to the gods, which were merely attributes of places and things. An intelligible law of causation was perforce as 'eternal' as the gods were supposed to be. Therefore, the teachings of these philosophers

---

[9] With few exceptions, the ancient Greek thinkers considered that the cosmos was bounded (limited). They could not endure the idea of 'infinity'. Indeed, the discovery of irrational numbers caused a furore.

diverged from religion and thus from poetry, but they retained the same *objectives*. Their conception of the observable world presupposed genealogy, bounded form, order, intelligibility and purpose – precisely the concerns of Homer and Hesiod, but now viewed naturalistically. Subsequently, Aristotle was to clarify those ideas and make them explicit, in prosaic rather than poetic language.

These pioneering attempts to explain the world naturalistically were as far removed from contemporaneous technology as any belief system of the ancient world. The reason is well known. Practical work, the use of existing technology, was the province of slaves and artisans. The pioneering philosophers belonged to the merchant and aristocratic classes, who *owned* slaves and *employed* artisans.[10] There was a wide social gulf between doers and thinkers, which precluded any effective linkage between belief and technology. Certainly there was nothing like the complex dialectic that exists in modern times. The apparent exception to this rule in the Greek world, Archimedes of Syracuse (he lived about three centuries after the pre-Socratic thinkers, roughly the time that has now elapsed since Isaac Newton's death), treated technological artefacts as phenomena to be accounted for by his mathematical models, along with the phenomena of the natural world. His theoretical interests outweighed his practical ones.

Matters were different in Western Europe as it emerged from the Middle Ages. Individuals could be both thinkers and doers. That may explain, in part, why knowledge and technology came to be so intimately linked in our society. But it also begs a question: *why* was there no social gulf between thinkers and doers in early modern Europe?

We shall leave that question aside until Chapter 5. In Chapter 3 we shall explore how naturalistic thought reached its culmination in Classical Athens, survey its subsequent development, and ask how it (or some of it) came to be preserved. The history we shall trace during the following four chapters helps us to understand how, when and why some of the deepest questions about the nature of science – and particularly the nature of biology – came to be asked.

---

[10] Aristotle, for example, put slaves below men and women and above animals on his 'ladder of nature'. This was tantamount to regarding them as an inferior species, precluding the possibility of useful dialogue. The 'ladder of nature' was the conceptual parent of the Neoplatonist 'Great Chain of Being', about which we shall say more in the next chapter.

Chapter 3
Classical Roots

Plato (c.427–c.347 BC)

Plato was born within living memory of the pioneers of Greek naturalistic thought. He was of noble birth, and by all accounts a gifted poet. In his youth he was strongly influenced by Socrates, so he developed a bent towards speculative thought and critical debate, particularly about matters of ethics and virtue. All Plato's work was written in the form of dialogues in which beliefs are subjected to intense critical analysis. The early dialogues focus on ethical issues and probably reflect the views of Socrates (who is usually given the major role in each debate). The middle and later dialogues reveal the mature Plato; they consider a wider range of issues, particularly the nature of knowledge and the nature of reality, with the character of Socrates often playing a lesser role.

In some of these mature dialogues, Plato scrutinised the beliefs of earlier naturalistic philosophers. He considered that the differences among them could be resolved if the claims and limitations of each were clearly specified. Searching for the requisite common ground made him focus on logic, precise definitions of terms and consistency of classification. It also made him willing to accept abstractions such as atoms and insist that mathematics – the most abstract possible way of thinking – was the basis of all understanding.[1] Plato's commitment to logic, mathematics and abstraction matured into a belief – the *theory of forms* – that perfect order was restricted to the world of ideas. The physical world, the world of material nature, contained only approximations to the perfect forms, which the mind alone could comprehend. Associated with this trend in Plato's thought was a hint of the notion of a single God: just as he sought the common basis of his predecessors' speculations about the nature of the world, so he sought the 'common ground' among the gods in which his society believed.

---

[1] Famously, Plato founded the Academy. Over the door of the building was inscribed the legend 'Let no one who is ignorant of mathematics enter here', testimony to Plato's belief in the fundamental importance of mathematics. The Academy was so called because it was apparently built on the site of a garden, belonging to one Akademos, that was sacred to the goddess of wisdom, Athene. It is interesting to reflect that our words 'academy' and 'academic' are derived from the name of a gardener, so the phrase 'gardens of academe' has genuine historical significance.

P.S. Agutter, D.N. Wheatley, *Thinking about Life*,
© Springer Science + Business Media B.V. 2008

He insisted that true knowledge and belief were the only reliable guides to action. In the dialogue *Meno*, for example, he put the following words into the mouth of Socrates[2]:

> Only these two things, knowledge and belief, guide correctly... The things that turn out right by some chance are not due to human guidance, but where there is correct human guidance it is due to two things, true belief or knowledge.

Plato inculcated these notions among his pupils, and more importantly, encouraged logical debate, but interpretations of his work differed and disagreements ensued. Six hundred years after Plato's death, in the 3rd century AD, these culminated in a mystical or spiritually-orientated derivative of Platonism that is now known as *Neoplatonism*. We shall say more about this later because it was to exert considerable influence on both Islamic and Christian mysticism, and thus on European thought.

Aristotle (384–322 BC)

Plato's most outstanding pupil, Aristotle, single-handedly developed the formal foundations of logic and the art of 'reviewing the literature'. Unlike his master, he became convinced that a perfect and essentially simple order underpinned the apparent chaos of material nature. The task of the philosopher, he thought, was *to uncover this underlying order in the observable world*; it was a mistake to seek order only in the realm of ideas. That is why some commentators consider Aristotle the fountainhead of scientific thought. However, because he was sceptical about Plato's theory of forms, he rejected mathematics as the basis of understanding, and he did not believe in atoms because they were not parts of *observed* nature. (In any case, he regarded time and space as relative to matter *and* as indefinitely divisible continua, so he considered atomism to be logically indefensible.) As far as we can tell from his surviving works – probably about a quarter of his total output – his account of the world was largely qualitative. Also, unlike modern science, it left no significant loose ends, and it was inherently static rather than progressive. Therefore, his thought differed in several important effects from what is now considered science (Fig. 3.1).

Aristotle was not a citizen of Athens so he was not allowed to teach within the city boundaries. He rented rooms in a gymnasium[3] complex to the east of the city, the Lyceum, where he and his successors worked. There he assembled what may have been the first library in Western history; it was dispersed after his death.

Aristotle's intellectual attainments have never been surpassed in the Western world. Apart from his ground-breaking achievements in the topics we now consider

---

[2] Translation by Grube GMA (1976) Hackett Publishing, Indianapolis, IN.

[3] This was more than a 'gymnasium' in our modern sense. It contained accommodation for philosophical debate, together with a number of temples, and was surrounded by gardens.

a

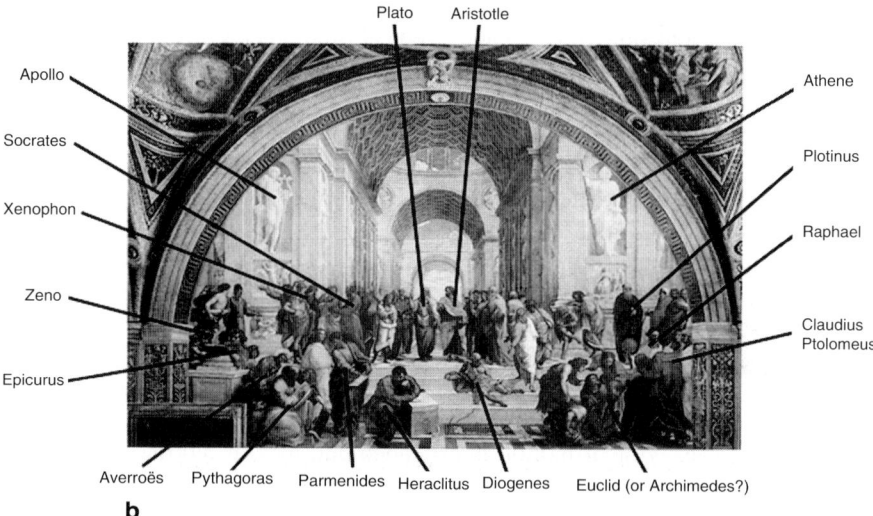

b

Fig. 3.1 The School of Athens (*Scuola di Atene*) painted by Raffaello Sanzio (Raphael, 1483–1520) *c*.1510 for the library of Pope Julius II and now in the Vatican. In the centre of the painting, Plato and Aristotle are shown walking together, locked in debate. The former is venerable and restrained, one hand pointing upward towards the heavens. The latter is young and vigorous, one hand pointing downwards towards the Earth. The smaller copy of the painting shown here has been annotated to identify the other figures whose identities are reasonably certain. The figure believed to represent Euclid, or perhaps Archimedes, is said to resemble Raphael's friend Donato Bramante, the Papal architect. The unidentified occupants of the picture probably include other members of the Pope's staff and/or Raphael's circle of friends. Raphael himself makes an appearance. It is interesting that the great 12th century Islamic philosopher Averroës (Ibn Rushd) is included in the 'School of Athens'. Averroës was the most celebrated commentator on the works of Aristotle and was instrumental in introducing Aristotelian thought into 12th and 13th century Europe (Chapter 4) 'science' (his *Physics* included meteorology and astronomy) and the foundations of logic, he wrote profound works on language, ethics, politics, aesthetics and metaphysics that remain influential today. He wrote a great deal about biology, as we shall see in Chapter 7. Many centuries after his death, his logic and physics came to be highly esteemed both in the Islamic world and in Christian Europe. Such was his influence on mediaeval Europe that many writers, following Thomas Aquinas, referred to him simply as 'The Philosopher'.

Aristotle's Physics

The cosmos according to Aristotle comprised two sharply-divided parts: the celestial and the terrestrial. In the former – that is, the sky above us – nothing was ever created or destroyed or changed in any quality. The only change in the celestial world was *natural motion*, which was perfectly circular, befitting the perfection of the celestial spheres. Aristotle asked what caused celestial objects – stars and planets – to move in their circular paths. '*Everything that moves,*' he wrote, '*is moved by something.*' The answer was that each planet[4] is attached to a transparent sphere and the sphere rotates. Each sphere is driven to rotate by the next adjacent sphere. The most distant sphere, the one to which all the stars were attached (hence the phrase 'fixed stars'), drove the others. This sphere, in turn, was driven by an unidentified but logically necessary *prime mover*.

In the terrestrial part of the cosmos, that is, everywhere below the sphere of the moon, 'natural motion' was not circular but linear. Drop a stone and it falls to the ground in a straight line. According to Aristotle, the stone has a natural tendency to move towards the centre of the Earth, i.e. the centre of the cosmos. In addition to natural motion there is violent motion. Throw a stone into the air – that is, impart violent motion to it – and it does not, initially, move towards the centre of the Earth. Gradually, however, violent motion is expended and natural motion resumes; the stone returns to the ground.

Why does a stone have a natural tendency to move downwards? Aristotle said that everything in the terrestrial world consisted of various mixtures of the four elements of Empedocles: earth, water, air and fire. The natural locations of these elements formed four concentric spheres: earth lowest, fire highest, water and air in

---

[4] The 'planets' were the Sun, the Moon, Mercury, Venus, Mars, Jupiter and Saturn. The presumption that the Earth was at the centre of the cosmos placed both Sun and Moon in the same category as the then-known planets. The other planets that we recognise today were not known to the ancient world because they cannot be seen without a telescope. 'Planet' literally means 'wanderer', and the name was applied because their observed paths through the heavens are more complicated than the simple circles described by the 'fixed stars'. According to later writers – most importantly Claudius Ptolemy – these complicated paths can be analysed as sets of superimposed circles, caused by smaller spheres (epicycles) rotating around the rims of larger ones (deferents).

between. An object consisting mainly of the element earth, such as a stone, tended towards its natural location, the sphere of earth, i.e. towards the centre of the world (and therefore of the cosmos). An object consisting mainly of the element fire, such as the smoke from a bonfire, would naturally rise towards the highest terrestrial sphere, immediately below the sphere of the moon.

Terrestrial objects were imperfect. Unlike entities in the celestial portion of the cosmos they were susceptible to changes other than motion: they could be created and destroyed, and they could change in size, shape and various qualities such as warm or cold, dry or moist. These qualities depended on the relative proportions of the four elements composing the object.

All changes, including motion (circular in the celestial part of the cosmos and linear in the terrestrial part), had causes. There were four kinds of cause: material, formal, efficient and final. The 'material cause' was the substance of the object involved in the process (ultimately, the elements it comprised). The 'formal cause' encompassed the shape and properties of that object. The 'efficient cause' was whatever or whoever imposed the form and properties. The 'final cause' was the object's purpose or natural place in the scheme of things. To understand any event or process in the observable world, a full account of its four causes was needed (Table 3.1).

This doctrine of four causes probably had its roots in Aristotelian biology (Chapter 7). Aristotle's studies of living things, and particularly of embryo development, profoundly influenced his account of the cosmos as a whole. His 'physics' was in that sense *organic*, in sharp contrast to modern physics. During embryo development, the uterus (formal cause) shapes the growing seed (material cause) towards the goal-state of the mature organism (final cause). Aristotle's accounts of all natural processes broadly conform to this account of embryogenesis. As a result, his world-view was entirely naturalistic but teleological rather than mechanistic. Every process must have a final cause, a 'goal,' just as embryo development does. He wrote[5]:

Table 3.1 Contrast between Aristotelian physics and Newtonian (classical) mechanics

Aristotelian	Newtonian
Celestial motion circular Terrestrial motion linear	All motion linear unless perturbed
Distinction between natural and violent motion	Distinction between constant velocity (or rest) and acceleration (external force applied)
Stones have natural tendency to fall towards the centre of the earth	Stones are accelerated towards the centre of the earth by the force of gravity
Everything that moves is moved by something	Constant velocity is maintained unless a force is applied
Motion and other changes have material, formal, efficient and final causes	Four-causes explanation abandoned
Conception essentially 'organic' and teleological	Conception entirely mechanistic

[5] Translated by Fine G, Irwin T (1991) Aristotle: Selections. Hackett Publishing, Indianapolis, IN.

How much, then, must the student of nature know about form and essence? As much, perhaps, as the doctor knows about sinews, or the smith about bronze – enough to know what something is for.

Athens after Plato and Aristotle

Both the Academy and the Lyceum continued to operate as centres of learning until the sack of Athens by Sulla in 86 BC. They were subsequently re-opened by teachers who claimed to maintain the traditions of their founders. The Lyceum probably ceased to operate around AD 300. The Academy survived until it was finally closed, together with all remaining schools of pagan philosophy, by the Roman Emperor Justinian I in AD 529.

Among the numerous philosophers (and philosophies) in Athens during this long period of history, we will mention only one because he became significant during the Scientific Revolution. Epicurus (341–270 BC), who like Aristotle was not a native Athenian, adopted the atomism of Democritus. He held that all objects and events, including humans, are nothing more than temporary physical interactions that occur by chance among countless indestructible particles. Because death is final annihilation, the best to be hoped for in human life is mental stability and the avoidance of pain and suffering; this negative kind of pleasure is the highest good. Epicurean philosophy was not accepted by the Christians and was considered disreputable throughout the European Middle Ages. Its reappraisal early in the 17th century coincided with the beginning of what we now call 'science'.

Alexandria

During Aristotle's lifetime, Alexander the Great conquered an empire that stretched from Macedonia and Greece to the north-west of the Indian subcontinent and encompassed Persia, Babylonia and Egypt. After Alexander's death the empire fragmented into independent units, many of which retained elements of Greek culture and learning for centuries. Foremost among these units was Egypt, which was conferred upon Alexander's half-brother Ptolemy. Ptolemy's successors, most of whom bore the same name, ruled Egypt from their new capital, Alexandria, until the country fell under Roman hegemony during the 1st century BC.

Ptolemy I established the celebrated Museum; the Royal Library of Alexandria[6] was probably completed during the reign of Ptolemy II. During its heyday, Alexandria was the intellectual hub of the Mediterranean world. The Library, which housed copies of almost every extant written work, attracted leading scholars from

---

[6]The library is said to have been organised initially by Demetrius of Phaleron, a pupil of Aristotle.

far and wide. It was there, for example, that Euclid (around 300 BC) wrote his compendious *Elements of Geometry* and founded a school of mathematics; Aristarchus of Samos (*c*.310–*c*.230 BC) proposed the notion of a heliocentric universe, calculated the distances from the Earth to the sun and the moon,[7] and inferred that the stars were very distant; Apollonius of Perga (*c*.262–*c*.190 BC) compiled pioneering studies of conic sections, optics and astronomy; Herophilos and Erasistratos ventured on vascular surgery during the 3rd century BC; and Claudius Ptolemy (*c*. AD 90–160) wrote his detailed account of the geocentric universe, the *Syntaxis*,[8] which provided far better predictions of celestial movements than any other work of antiquity. Home to most of the Classical world's contributions to knowledge, the Library was damaged by fire on at least one occasion, and finally destroyed – so it is alleged – by arson. The circumstances have been much debated, but the culprits were probably Christians during the 4th century AD.

Politically, Alexandria was the capital of Egypt, but it retained much of the character of the Greek city states. Its élite remained linguistically and culturally Greek, though they did assimilate some aspects of Egyptian lifestyle (e.g. clothing and religion). The rest of Egypt remained resolutely non-Greek. Moreover, Alexandria was a major importer and exporter of goods and had trade links that extended not only throughout the Mediterranean world but over the Middle East and as far as India. It therefore had many of the characteristics to which we attributed the origin of naturalistic thought (Chapter 2) – and of course it inherited and preserved a powerful tradition of such thought. Its learning went into relative decline after 30 BC when Egypt was assimilated into the highly centralised and increasingly bureaucratic Roman Empire.[9]

Some significant additions to the corpus of learning continued in Alexandria during the early centuries AD, for example in the mathematics of Diophantus, but Romanisation had unexpected effects. Rich Romans employed Greek slaves as teachers, scribes and accountants. These highly educated Greeks were thus obliged to mix socially with manual workers, and there was occasional cross-fertilisation of skills – the practical with the intellectual. Thus, during the Roman and early Christian period, we find instances of technology inspired by ideas, and vice versa. In the 1st century

---

[7] These calculations presumed a geocentric universe; Aristotle and virtually all other authorities until the time of Copernicus believed that the Earth was at the centre of the universe. It is not clear whether Aristarchus took his heliocentric hypothesis seriously.

[8] The *Syntaxis*, or *Mathematical Compilation*, entered mediaeval Europe via an Arabic translation and commentary, the *Almagest*. This work remained the major astronomical and astrological authority until well into the 16th century. Claudius Ptolemy was also a renowned geographer and also wrote works on music theory and optics; his geographical treatises served as a guide to Columbus and others.

[9] Alexandria did not have a monopoly on intellectual endeavour. Another great library existed at Pergamum, a Greek city located in modern Turkey some 16 miles from the Aegean coast. Also, Archimedes of Syracuse (*c*.287–*c*.212 BC), the greatest mathematician of antiquity and a pioneer in mechanics and practical astronomy, seems to have spent most of his active life in Sicily. But Alexandria was unquestionably the successor to Athens as the centre of learning in the Classical world.

AD, Hero of Alexandria combined his mastery of geometry with practical knowledge to devise a mechanical fountain and a steam-driven engine, the aeolipile; he was also a skilled draftsman and land surveyor and made contributions to mechanics and optics. He established a technical school – in effect, a 1st century polytechnic. Thanks to such initiatives, the blending of learning with practical skill survived in Egypt, and perhaps other parts of the empire, until after the fall of Rome.

Concurrently, however, the reduction of Alexandria's status from (in effect) independent city-state to provincial Roman town seems to have facilitated a reintroduction of supernaturalistic thought into its culture.

Mystical and Religious Thought in Alexandria

During the early centuries of the Christian era, Alexandria was the birth-place not only of many seminal works of Judaic and Christian theology but also of independent quasi-religious mystical schools such as Hermeticism and what is now termed Neoplatonism.

Neoplatonism denotes a school of philosophy that originated with Plotinus (AD 205–270).[10] It fused many strands of earlier Greek philosophy, notably Aristotelian logic, Platonic idealism and Pythagoreanism. Its teachings were highly complex. Essentially, it held that the material universe was a product ('emanation') of the divine, ultimately of the unknowable One; that the divine acted in the world via a succession of lesser spiritual beings; and that the individual soul's quest for the Good was a quest for reunion with the One. Different varieties of Neoplatonism were taught in Alexandria, Rome, Athens and Damascus until the 5th or 6th centuries. Later, it was to influence the teachings of some Islamic Sufis and some varieties of Buddhism as well as European Christian mysticism. The great thinkers of late mediaeval Europe were conversant with it.

We owe much of our knowledge of Plotinus to a biographical account by one of his leading pupils, Porphyry (*c.* AD 233–305), who was educated in Athens and became a disciple of Plotinus in Rome during the 260s. Porphyry simplified his master's teachings, which made him more congenial to the practically-minded Romans, but two aspects of his thought were especially significant for later developments. First, he began a tradition that considered the relationship between the One and the rest of the spiritual hierarchy in terms of a triad (abiding, procession and reversion), which was loosely based on an argument in a late Plato dialogue, *Timaeus*, and seems to have influenced the Christian doctrine of the Trinity. Second, he assimilated astrology into his philosophy, making considerable use of the work of Claudius Ptolemy. In the long term, the relationship between astrology and Christian thought proved delicate though durable. However, astrology became a serious intellectual pursuit in both the Islamic and Christian worlds.

---

[10] The philosophy of Plotinus is said to have been influenced by his teacher, an Alexandrian dock worker called Ammonius Saccus, about whom little or nothing is known.

Details of the alleged interrelationships between the various levels of the divine and the various levels of physical reality, from inanimate objects to humans, ultimately gave rise in Christian Europe to the concept of the Great Chain of Being. This was a linear hierarchy extending from inanimate matter to God, and was considered a development of Aristotle's Ladder of Nature.

Hermeticism took its name from the Greek god of wisdom, Hermes, who was identified with his Egyptian counterpart, Thoth. It was based on texts attributed to an Egyptian priest and sage, Hermes Trismegistos, but probably written in the 1st century AD (Plotinus and Porphyry were familiar with them). Essentially, Hermeticism is pantheistic and indicates that understanding and spiritual fulfilment can be attained by alchemy, astrology and theurgy. 'Alchemy' originally meant the conversion of a base person into a spiritual one and was only later applied to the transmutation of metals. It required the use of a 'tincture', the formulation of which was not specified,[11] though various obscure authorities were said to have procured it. 'Theurgy' was divine magic requiring alliance with divine spirits. Hermeticism had a clear morality and many of its practitioners have found it possible to connect their beliefs with the mystical aspects of a major religion – Buddhism, Judaism, Christianity, Islam – but it was officially condemned by most Christians and Muslims and became part of an occult underground in mediaeval Europe. It influenced a number of significant contributors to the Scientific Revolution.

The Dispersal and Reunion of Classical Learning

Despite the destruction of the Alexandria Library, a significant amount of Classical learning was preserved after the 4th century AD. Our concern here is to trace its subsequent fate and to explain how it reached mediaeval Europe (see Fig. 3.2 for summary).

Roman learning. It has been said that Rome conquered Greece, but Greece educated Rome. Greek learning was certainly esteemed during the height of the Roman Empire, but the Romans made few significant additions to the achievements of Classical Athens and Alexandria; they excelled in engineering and practical arts, not in novel contributions to philosophy and naturalistic thought. Even such luminaries as the elder Pliny mixed the fantastic with the real, apparently indiscriminately.[12] However, after Christianity was adopted as the state religion in the 4th century, the preservation of Classical learning in both the Eastern and Western Empires proved lastingly important.

---

[11] 'Tincture' is a more or less literal English translation of the original Greek χυριον (*chirion*). The word was adapted into Arabic by the addition of a definite article and the omission of the inflected ending – *al-chir*, hence *al-iksir*, and this was subsequently rendered in Latin translations from the Arabic as *elixir*.

[12] Nevertheless, Pliny could be a fine observer. He died during the eruption of Vesuvius that destroyed Pompeii, having observed the catastrophe from across the bay; his last written words are now regarded as the first accurate and detailed description of a pyroclastic flow.

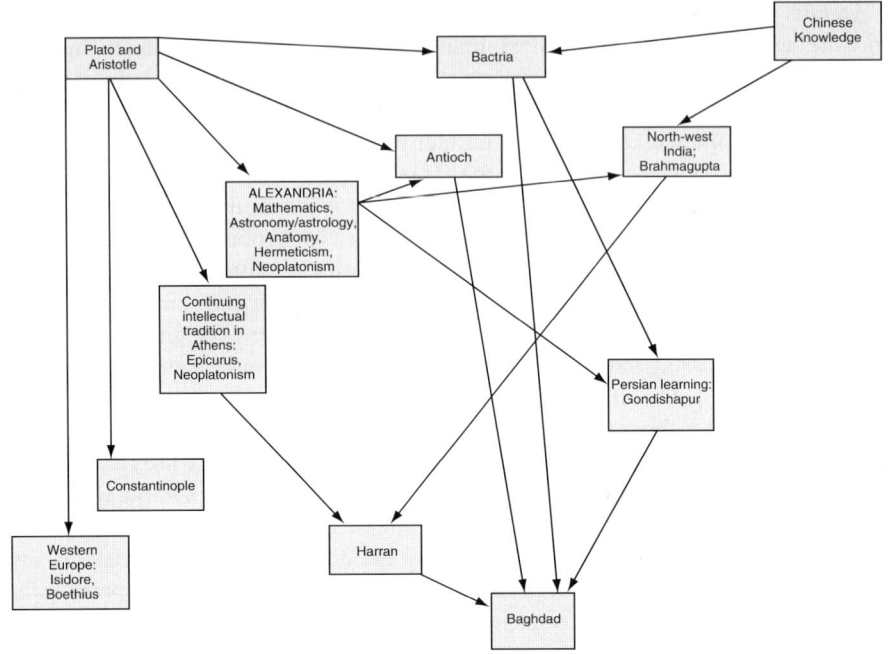

Fig. 3.2 Summary of the routes through which Classical Athenian learning reached Western Europe, Constantinople and Baghdad. The first of these 'destinations' received only fragments of Aristotle and Euclid, and a single complete Plato dialogue, *Timaeus*. Much more material was preserved in Constantinople and Baghdad

Classical texts were preserved in the capital of the Eastern (Byzantine) Empire, Constantinople. They included many complete (edited but largely uncorrupted) works of Aristotle and Plato, as well as manuscripts by Archimedes, and by Aristarchus and other scholars associated with Alexandria. Copies of these texts, in the original Greek, entered Western Europe from the 13th century onwards, and especially in the 15th century, after Constantinople fell to the Ottoman Turks and was renamed Istanbul.

As the Western Roman Empire was dying, Classical texts were translated into Latin, mainly by Boethius (*c*.475–*c*.524), who had been educated at the Academy in Athens before taking up a government post in Italy. Portions of Aristotle's work, together with fragments of Classical mathematics (including Euclidian geometry), came to be known in Dark Age Europe thanks to Boethius. Isidore of Seville (560–636) pioneered the establishment of cathedral schools and the teaching of the liberal arts[13] and some elements of law, medicine and other disciplines, in addition to

---

[13] The liberal arts comprised the trivium (grammar, rhetoric and logic) and the quadrivium (arithmetic, geometry, astronomy and music). They formed the basis of university education throughout the European Middle Ages. Though pioneered by Isidore, this institution was truly established by Alcuin of York (*c*.735–*c*.804), who designed the curriculum for the palace school founded by Charlemagne and was emulated thereafter.

theology. Boethius and Isidore were both committed Christians and were much influenced by Neoplatonism.

Remnants of Alexander's Empire. Bactria (modern East Turkmenistan and Northern Afghanistan), possibly the birthplace of Zoroaster, had been part of the Persian Empire. It became independent Greek state after Alexander's death; Greek classics were translated there into a dialect of Iranian. During its height it established contact with China and later became part of silk route. Subsequently it fell under the control of various outside rulers and was finally conquered by Arabs during the 7th century.

Antioch was founded around 300 BC by Seleukos, another of Alexander's generals, and became a centre of learning and later one of the four greatest cities of the Roman Empire. Highly cosmopolitan and a melting-pot of cultures and religions, it was dubbed the 'Athens of the Near East'. It was also an important centre for Christianity during the 1st century AD. Greek classics were translated in Antioch into several languages, notably Syriac. Although the city was destroyed by earthquake and famine in the 6th century, many of these translations survived.

Alexander's Empire had only just reached what is now the north-west of India, but trade links with Alexandria had carried Classical learning to this area. Brahmagupta (AD 589–668) had a clearly Aristotelian view of nature but nevertheless made important advances in mathematics and astronomy. He was head of the astronomical observatory at Ujjain, some 250 miles north-west of Nagpur, and was one of several Indian theorists to consider a heliocentric model of the universe. His *Brahmasphutasiddhanta* used Hindu-Arabic numbers and is the first known text to treat zero as a number in its own right. Brahmagupta pioneered *kuttaka* (which we now call algebra); he discovered the quadratic formula and found solutions to a several problematic equations. He applied kuttaka to astronomy, calculating conjunctions, times of the rising and setting of stars, eclipses, and many other phenomena.

Refugees from the Academy in Athens. When Justinian closed the schools of philosophy in 529, the remaining teachers at the Academy moved to Harran, which now lies in Turkey close to the Syrian border. They took with them a number of works of astronomy, physics, medicine and philosophy, dominated by Neoplatonism, and formed an 'Academy in exile'. This survived more or less unmolested for almost 300 years.

Gondishapur. What may be considered the world's first cosmopolitan university and hospital were built in Gondishapur, south-western Persia, in the 3rd century AD. Although the knowledge base in this university was indigenously Persian, particularly in the fields of anatomy and surgery,[14] many Roman texts were transferred there and translated into Pahlavi. Greek, Syriac, Indian, Greek and Alexandrian learning were mingled with Iranian knowledge, and philosophy, mathematics,

---

[14] These disciplines were very advanced because of the long tradition of human dissection in Persia. It is plausible that Zoroastrian influence had encouraged scholars to see the human body as a microcosm of the universe as a whole – an idea that was broadly consistent with Aristotle's 'organic' conception of the cosmos. Books from Gondishapur, translated into Arabic, entered Europe during the Crusades and were translated again into Latin, then later into French and English.

theology, music, political sciences and agriculture were taught as well as medicine. The Islamic conquest of Persia (AD 637–651) ended the dominance of Gondishapur; Baghdad succeeded it as a centre of learning.

The Muslim Empire. The spread of Islam during the 7th century AD and afterwards was instrumental in bringing together these scattered and much-translated classics – including the mystical writings of Hermeticists and Neoplatonists – and the technological innovations of Roman Alexandria. During the late 8th and 9th centuries, Harran was a centre for translating works of astronomy, philosophy, natural sciences and medicine from Greek to Arabic, bringing the knowledge of the classical world to the emerging Arabic-speaking civilization. Many important scholars of natural science, astronomy and medicine originated from Harran, possibly including the alchemist Geber (see Chapter 4).

This development was an inspiration. Caliph Al-Mansur (712–775), who founded Baghdad as a great centre of learning, invited the scholar Kankah from Ujjain to explain Indian mathematics and astronomy. At the Caliph's behest, Al-Fazari translated the *Brahmasphutasiddhanta* into Arabic. Syriac texts from Antioch, and the Iranian translations of classics in Bactria, were likewise translated into Arabic. What followed was another great flowering of learning, first at the House of Wisdom in Baghdad and subsequently also in other great Muslim centres, notably Cairo, Damascus and Córdoba.

The impact of this development on mediaeval Europe has sometimes been underestimated, but it was decisive. Properly appreciated, it shows that a *continuous* (if rather convoluted) *tradition of thought* connects modern science to the Athens of Plato and Aristotle – though this tradition did not regain its wholly naturalistic character until after the 16th century (Fig. 3.3).

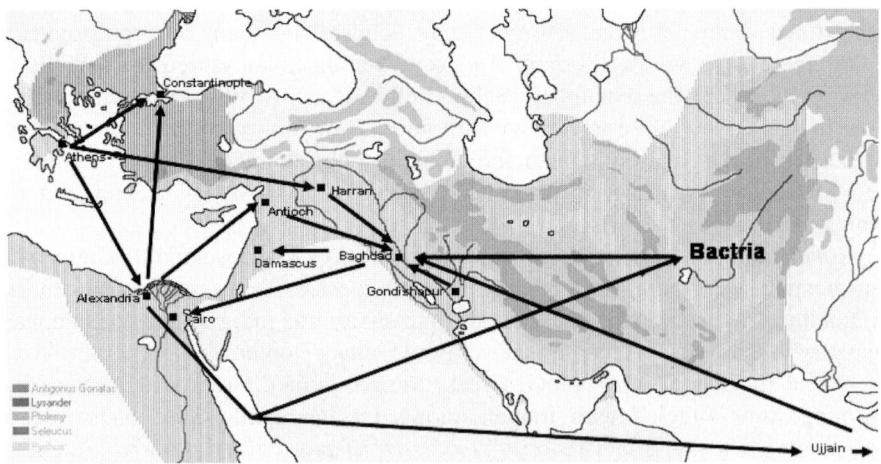

Fig. 3.3 Outline map of the routes though which fragments of Classical learning were distributed from Athens and Alexandria, reaching Constantinople in the early centuries AD and Baghdad in the 8th century. From Baghdad, learning passed to other Islamic centres including Cairo and Damascus. This portrayal of the major routes is superimposed on a map of Alexander the Great's empire, illustrating the relevance of the Macedonian conquest to the subsequent spread of learning

Chapter 4
Mediaeval Views of the World

Traditional histories of science acknowledge that the Arabs preserved much Classical learning. Sometimes they name a few leading scholars of the Muslim world in the 8th–12th centuries, usually in Latinised forms: Al-Jabr (Geber), Al-Khwarizmi, Ar-Razi (Rhazes), Ibn Sina (Avicenna), Ibn Rushd (Averroës). But such accounts are usually intended as a prelude to the 'real story', i.e. European developments after the 16th century (Chapter 1). They seldom consider that the prelude is, or should be, a story in itself: yet the European developments that culminated in modern science were presaged by Muslim contributions.

What were these contributions, how did the Muslim world come to make them, and why have they been relegated to little more than footnotes in most histories of science?

The Fostering of Knowledge in the Muslim World

Great religions are susceptible to different interpretations, which lead to sectarian divisions. The 'liberal' interpretation of Islam during the early caliphate of Baghdad was sustained for several centuries, though sometimes opposed. The Koran's injunction to 'read' was taken literally. Observation, reason and learning were not seen as inconsistent with Islam. On the contrary, knowledge and understanding of the world increased one's appreciation of the greatness of God, so learning was actively fostered. Together with the availability of classical texts, this partly accounted for the Islamic 'golden age'. But were the alleged conditions for naturalistic thought (Chapter 2) met?

To some extent, they were. First, the Muslim empire was very extensive and comprised many peoples, traditions and religions. Polytheisms were outlawed, but other monotheisms (Judaism and Christianity) were accepted.[1] Irrespective of

---

[1] Followers of other monotheisms were, of course, encouraged to convert to Islam; one practical and often effective form of encouragement was to tax Jews and Christians more heavily than Muslims.

P.S. Agutter, D.N. Wheatley, *Thinking about Life*,
© Springer Science+Business Media B.V. 2008

religion and history, contrasting cultures aroused interest. Scholars studied and recorded them in detail, opening their minds to critical comparison of ideas. Remarkable works of geography and ethnography resulted.

Second, although the empire was unified in terms of religion, it was not bureaucratic or administratively centralised. There was considerable regional autonomy. A large degree of cultural and political independence survived, and there was an element of competition (which sometimes flared into military conflict). There was a level of sophistication, even opulence, seldom approached in contemporaneous Europe.

Third, although merchants were not universally respected, they enjoyed autonomy. The accumulation of wealth through honest trading was considered God's reward for a good life. Trade fosters the exchange of ideas and encourages innovation. Wealth was created in a mostly stable and reasonably secure environment.

Fourth, the Arabs used alphabetical writing, and the vocabulary of the language expanded as ideas were adopted from Classical, Indian and other sources. This encouraged abstract thought.

However, Islamic thought in this era was hardly 'naturalistic'. Like the Scholastic tradition in 12th–16th century Europe (see below), it was dominated by commitment to a powerful monotheism – God was never far from the topic under discussion. As the great 12th century philosopher Averroës wrote[2]:

> We maintain that the business of philosophy is nothing other than to look into creation and to ponder over it in order to be guided to the Creator – in other words, to look into the meaning of existence.

Thought may have been fairly free, but thinking was dominated by religion, as was the case in Europe.

The Greek classics, Hermetic writings and works of Roman-Alexandrian technologists were translated and studied, leading to numerous developments in mathematics, alchemy and chemistry, astrology and astronomy. Perhaps most remarkable – and most influential – were the innovations in anatomy, physiology and various aspects of medicine (Table 4.1 summarises some precedents for later European advances). Iran was particularly prominent in medicine, building on her long tradition of anatomy and surgery, which had culminated in Gondishapur (Chapter 3).

Advances in Knowledge during the Islamic Golden Age

There were similarities between the Islamic revival of learning after the 8th century and the European revival after the 16th. Both were marked by an excited response to Classical works; there were advances in mathematics and in medicine and almost

---

[2] Averroës (c.1190) On the Harmony of Religions and Philosophy (Kitab fasl al-maqal). Less than a century later, Roger Bacon wrote '*It is only the fool that chooses reason over faith or faith over reason for God is the giver of both*'; and Aquinas (Summa Theologica, part 1, first article) wrote '*It was necessary for man's salvation that there should be a knowledge revealed by God, besides the philosophical sciences investigated by human reason*'. The similarity of viewpoint among these great Muslim and Christian thinkers needs no emphasis.

Table 4.1 Muslim precedents for aspects of European science

Topic	European pioneer(s)	Muslim precedent(s)	Comments
General scientific ideas			
Experimental approach to science	Roger Bacon (c.1214–c.1292); Galileo Galilei (1564–1642)	Jabir (c.721–c.815); Ar-Razi (864–930); Al-Biruni (973–1048)	
Indestructibility of matter	Antoine Lavoisier (1743–1794)	Al-Biruni (973–1048)	
Atomism	Pierre Gassendi (1592–1655)	Ar-Razi (864–930)	Revival of teachings of Epicurus
Disciplines			
Astronomy	Alphonsine tables (13th century)	Toledo tables (9th century and following)	Planetary motion revised from Claudius Ptolemy
Chemistry	Robert Boyle (1627–1691)	Jabir (c.721–c.815); Ar-Razi (864–930)	Critical approach to alchemy
Geology/geomorphology	Nicolas Desmarest (1725–1815); James Hutton (1726–1797)	Ibn Sina (980–1037); Al-Biruni (973–1048)	Formation of mountains and valleys
Palaeontology	Leonardo da Vinci (1452–1519)	Al-Biruni (973–1048)	Explanation of fossils on mountains
Mathematics			
Trigonometry	Mertonian Calculators, Oxford, 14th century	Al-Battani (c.853–c.929)	Used trigonometric relationships to correct Claudius Ptolemy
Algebra	François Viète (1540–1603)	Al-Khowarizmi (c.778–c.850); various 9th century mathematicians	First use of algebraic symbols; derived from Brahmagupta's *kuttaka*
Cubic equations	Niccolò Tartaglia (1500–1557)	Various mathematicians early 10th century	Derived from Brahmagupta's *kuttaka*
Decimal fractions	Simon Stevin (1548–1620)	Al-Kashi (early 15th century)	First use of decimal point
Negative numbers	Geronimo Cardano (1501–1576)	Various mathematicians 12th century	
Logarithms	John Napier (1550–1617)	Mathematicians 13th century	
Binomial theorem	Isaac Newton	10th century mathematicians	

(continued)

Table 4.1 (continued)

Topic	European pioneer(s)	Muslim precedent(s)	Comments
Medicine and related sciences			
Scientific surgery	Ambroise Pare (1510–1590)	Az-Zahrawi (936–1013)	Ligation of arteries; wound sealing
Hospital care for the insane	Philippe Pinel (1745–1826) in 1793	11th century and after	
Anaesthesia	Opium extracts: Paracelsus (1493–1541); Inhalation anaesthesia: W.T.G. Morton (1819–1868)	Opium: Ibn Sina (980–1037), who took the idea from the Greeks; Inhalation anaesthesia: Az-Zahrawi (936–1013), Ibn Zuhr (1091–1161)	Narcotic-soaked sponges placed over the patient's face were used in 13th century Islamic Spain.
Antisepsis	Joseph Lister (1827–1912)	Surgeons in Islamic Spain	Used alcohol as antiseptic and promoted post-operative hygiene
First pharmacopoeia	Germany, mid 16th century	Ibn Baytar (c. 1190–1258)	
Pharmacology	Paracelsus (1493–1541); main advances late 19th and 20th centuries	9th century physicians and alchemists/chemists	Numerous plant extracts examined
Drug treatment of infections	Paul Ehrlich (1854–1915)	Ar-Razi (864-930) and others	Sulphur and mercurial treatments
Pulmonary circulation	Michael Servetus (c.1510–c.1553); Realdo Colombo (c.1516–c.1559)	Ibn al-Nafs (1213–1288)	
Pathology	Giovanni Morgagni (1682–1771)	Ibn Zuhr (1091–1161); Az-Zahrawi (936–1013); other surgeons	Various circulatory and thoracic diseases were distinguished; abdominal cancers recognised
Embryology		Ibn al-Quff (1233–1289)	Note Aristotle's seminal work
Scientific instruments			
Glass mirrors	Venetians, 1291	Syrians 9th century; Islamic Spain	
Lenses and prisms	Witelo (born c.1230; Roger Bacon; Johannes Kepler (1571–1630); Isaac Newton (1642–1727).	Al-Haytham (965–1039)	
Pendulum	Galileo Galilei (1594–1642)	Ibn Yunus al-Masri (c.950–c.1009), Egypt	

Technology			
Concept of human flight	Roger Bacon (c.1214–c.1292); Leonardo da Vinci (1452–1519)	Ibn Firnas (810–887)	Actual flight achieved in Islamic Spain in the 870s
Mechanical clock	Milan 1335	Ibn Firnas (810–887)	Other scholars in Islamic Spain had astronomical clocks
Printing press	Johannes Gutenberg (c.1400–c.1468), 1454	Islamic Spain early 14th century	Moveable brass type
Gunpowder	Roger Bacon	Probably first devised by Muslim chemists	Chinese discovered salpeter (potassium nitrate); may have acquired gunpowder by trade with Muslim world
Compass/navigation	Alexander Neckam (1157–1217)	11th century or earlier	Probably first developed in China

Many earlier inventions and discoveries (e.g. antibiotics) are reasonably attributed to the Chinese. The Indians, too, had very old astronomical devices, though some may have originated from Persia. We do not disregard these achievements, but their influence on the development of western science, if any, was indirect, and was mainly transmitted through the Muslim world. As we noted in Chapter 1, no pattern of thought resembling modern science ever arose except in western Europe at the end of the mediaeval period.

all new biological knowledge came from medicine, though there was biological speculation aplenty. There were also differences. The so-called Scientific Revolution in Europe was centred on mechanics, the construction of an anti-Aristotelian physics, but mechanics was less prominent for most Muslim scholars. On the other hand, Muslim thinkers made huge advances in alchemy and chemistry, while chemistry did not become a fully-fledged discipline in Europe until the end of the 18th century. For example, Abu Musa Jābr ibn Hayyān (Geber; *c*.721–*c*.815), court alchemist to the fifth caliph, Harun al-Rashid[3] (763–809), described techniques of distillation, crystallization, sublimation and evaporation, manufactured acetic acid and various mineral acids including aqua regia, and anticipated the law of constant proportions. In these respects he was 'a chemist' rather than 'an alchemist', though his writings subsequently became the standard texts of European alchemy.[4] He also wrote works about medicine and astronomy. Thanks largely to Geber's pioneering endeavours, Muslim scholars went on to made striking advances in chemistry and especially in its medical applications[5] (Table 4.1).

One such scholar, the polymath Ar-Razi (Rhazes, 864–930), reasoned that if time and space are regarded as independent of matter, atomism is logically tenable. This single example shows that Islamic scholars were not content simply to translate, read and believe Aristotle and other classical authors; they *evaluated* them and were prepared to disagree. Rhazes was also critical of Galen and corrected several aspects of Galenic anatomy (see Chapter 7). He may have recognised the function of venous valves some 700 years before Harvey.[6] By the same token, these scholars did not merely translate the mathematical works of Classical Greece and India, but developed them in new ways. For example, Thabit bin Qurrah (863–901) and Abu'l Wafa (*c*.940–*c*.997) discovered that algebraic methods could be used to solve geometrical problems, and this led to advances in trigonometry and its application

---

[3] Harun al-Rashid, the fifth caliph, divided the empire among semi-autonomous rulers, which led to long-term weakening and disintegration. But he was a great champion of learning, which embellished his famously opulent court.

[4] Unfortunately, much of his (surviving) work was written in a deliberately obscure coded language; therefore, this great scholar was to lend his name to the word 'gibberish'. He paid explicit tribute to the Hermetic writings and the numerology of Neoplatonism. But there was nothing obscure about his insistence on practical experimentation: *'The first essential in chemistry is that you should perform practical work and conduct experiments, for he who does not perform practical work or make experiments will never attain the least degree of mastery'*. Many centuries would pass before this sentiment was echoed in Europe.

[5] Moderately pure alcohol had probably been distilled by the Muslim chemists by the 10th century; it was used as an antiseptic in Islamic Spain during the 11th century. Most Western histories, however, tell us that alcohol was first purified by the Spanish alchemist Arnau de Villanova (*c*.1235–*c*.1313) early in the 14th century; Villanova is known to have translated a number of Arabic treatises.

[6] When Harvey was a student at the medical school in Padua in the late 16th century, the standard teaching texts were those of Avicenna and Rhazes. Harvey acknowledged neither scholar in his *magnum opus*, though he was generous in his praise of his mentor, Fabricius, and even of his rivals such as Riolanus.

to astronomy (measuring the declinations of stars). Not until the 14th century was trigonometry developed in Europe, and not until the 17th was the algebraic approach to geometry fully appreciated, initially by Descartes.

Most Western histories mention the *Continens Liber* (*al-Hawi*) by Rhazes[7] and the *Canon of Medicine* by Avicenna (Ibn Sina), which became the principal medical texts of mediaeval Europe. However, Avicenna's illustrious contemporary and fellow-countryman al-Biruni is scarcely known in the west. Avicenna was a child prodigy and workaholic who died before he was 60; al-Biruni was no less of a polymath. Apart from his encyclopaedic works on geography, geology, ethnography, religion and history, particularly of India, he made remarkable contributions in mathematics (summation of series, algebraic equations, irrational numbers, combinatorial analysis), astronomy (the galaxy consists of stars, the solar system may be heliocentric, the earth rotates) and physics (light has a high but finite velocity, acceleration is non-uniform motion). He also constructed an astrolabe, a plani-sphere and a mechanical lunisolar calendar computer. He engaged in a long-running debate with Avicenna about Aristotle's physics, particularly the belief that circular motion was 'natural' in the celestial cosmos (which Avicenna accepted and al-Biruni challenged). In scope and originality, al-Biruni compares with Leonardo da Vinci.

In biology, Al-Jahiz (*c*.781–*c*.868) introduced the concept of food chains, which had no known precedent in Greek or Persian thought, and also proposed a scheme of animal evolution that entailed natural selection, environmental determinism and (possibly) the inheritance of acquired characteristics. A more detailed idea of evo-lution was proposed by Ibn Miskawayh (932–1030), who proposed that God had invested matter with energy for development from vapour to water to minerals, and from thence to 'lower' animals and up the great chain of being to apes and humans. Men developing the gift of prophecy further evolved into angels. These general notions of evolution, implying a continuum between the non-living and the living (an 'organic' conception of matter), were debated among other Muslim scholars and became known in Europe through Latin translations.

Many other great scholars of Islam's 'golden age' have been similarly overlooked in the west (Table 4.1).

Averroës and Algazel

Al-Ghazali (Algazel; 1058–1111) was an outstanding theologian, philosopher and mystic who played a large part in curtailing the freedom of thought that had char-acterised the early centuries of Islam. Prominent among his many books was *The*

---

[7] Rhazes distinguished clearly between smallpox and measles, criticised the Roman anatomist Claudius Galen (*c*.129–*c*.216) on several points and took a Hippocratic attitude to medical prac-tice. Avicenna distinguished pleurisy from pulmonary tuberculosis and recognised that infections can be spread by water or soil. These two intellectual giants, both Iranian, were many centuries in advance of their European counterparts.

Incoherence of the Philosophers, in which he rejected the giants of classical antiquity, including Aristotle, as 'unbelievers' and 'corrupters of Islamic faith'. He introduced a sceptical view of knowledge that was later reflected, without acknowledgment, in the writings of Descartes and more particularly of Berkeley and Hume. In particular, he challenged the notion of material causality. He regarded 'causal' conjunctions as evidence for the immediate effects of the will of God, much as Berkeley was to do 600 years later. Algazel's influence was evident not only in much subsequent Muslim thought but also in mediaeval Christian Europe. His writings were widely studied, especially at the University of Paris, during the 12th and 13th centuries.

A generation later, Algazel's arguments were countered by 'the Arabic Aristotle', Ibn Rushd (Averroës; 1126–1198), whose work was a major influence on subsequent European thought. In *The Incoherence of Incoherence*, Averroës deployed his detailed understanding of Islamic theology and law, and his remarkable knowledge of Aristotle, to defend the writers of classical antiquity against Algazel's critique. He reaffirmed that critical philosophical inquiry is consistent with religious belief. He argued that we have two ways of knowing Truth: faith-based religion, which cannot be tested; and (for those with sufficient ability) reason-based philosophy. This notion of 'two truths' was to have a profound influence in mediaeval Europe. It came to legitimise the separation between Church and State that underpinned the process of *secularisation*. As we shall see, European secularisation was a prerequisite for the emergence of modern science.

Averroës used a subtle rhetorical device in his highly influential commentaries on Plato and Aristotle. He deployed a style that was customarily used for commentaries on the Koran: a simple overview followed by an intermediate comment and then by critical evaluation. Presumably his aim was to disarm those who, like Algazel, may have considered his work 'anti-Islamic'. He used a similar approach in a commentary on Avicenna's *Canon of Medicine* and an important compilation of Galen. Latin translations of these commentaries were instrumental in introducing Aristotle's (and to a lesser extent Plato's) works to the new universities of Europe after the early 1100s. So great was the influence of Averroës on European Scholasticism that Aquinas and others referred to him simply as 'the commentator', just as they referred to Aristotle as 'the philosopher'.

However, Averroës was much more than an editor and commentator – he was a highly original thinker. His studies of geography led him to predict the existence of a continent beyond the Atlantic Ocean. In astronomy, he gave the first description of sunspots and was critical of the Ptolemaic system:

> To assert the existence of an eccentric sphere or an epicyclic sphere is contrary to nature… The astronomy of our time offers no truth, but only agrees with the calculations and not with what exists.[8]

Most interesting, perhaps, was his contribution to physics. He developed the suggestion (attributed to his short-lived near-contemporary in Islamic Spain, Ibn

---

[8] Quoted by Gingerich O (1986) Islamic astronomy. *Sci Am* 254 (10):74.

Bajjah) that uniform motion continues in the absence of an applied force, an idea later to be formalised in Newton's first law of motion (cf. Table 3.1). Averroës also offered the first formal definition and measurement of 'force' as 'the rate at which work is done in changing the kinetics of a material body'. This is closer to what we would now define as 'power', but it was a remarkable contribution and led directly to the mathematical understanding of mechanics that blossomed in Europe a few centuries later.

A contemporary of Averroës, the great Jewish rabbinical teacher Maimonides (1135–1204), recorded and practised the medical teachings of ibn Rushd and Ar-Razi. Some of his writing was Neoplatonist in character. He was extremely sceptical about astrology, which he had studied in detail. Above all, however, he was another influential Aristotelian. His *Guide to the Perplexed*, which sought to reconcile Aristotelian philosophy with the Torah, became well known among European scholars in the 13th and 14th centuries.

Contemporaneous Changes in Europe

While the Muslim world grew sophisticated, wealthy, reasonably stable, and more or less unified, Europe for the most part remained poor, fragmented and subject to recurrent barbarian attacks that precluded stability. Some learning survived (largely in monasteries), thanks to the works of Boethius and Isidore, but this was limited and there were few advances in understanding of the natural world. Alcuin's educational initiative at the court of Charlemagne, and isolated champions of learning such as Bede[9] and Alfred, brought only feeble beams of light to a dark age. Pockets of literacy survived in Italy, and Charlemagne exploited them for his court's use; but ignorance, barbarism and superstition remained ubiquitous.

Just as Charlemagne's court in Aachen had institutionalised the teaching of the seven liberal arts (Chapter 3) and fostered learning,[10] so it instituted the mode of socio-economic organisation known as feudalism: 'no land without its lord, no lord without land'; everyone except the king himself 'belonged' to a feudal superior. The system helped to stabilise society and agricultural production, as well as existing technology allowed, and it organised the provision of troops and weapons to raise armies. It survived the collapse of Charlemagne's empire and became universal throughout Europe for the better part of a millennium. The feudal hierarchy was

---

[9] Bede (*c.*672–*c.*735) wrote *On the Nature of Things* and several books about mathematics and astronomy, including *On the Reckoning of Time*. He made original discoveries about the nature of the tides. His works became required subjects of study for the clergy and significantly influenced early medieval knowledge of the natural world.

[10] Harun-al-Rashid, Charlemagne's contemporary, sent him gifts that included an elephant. Nevertheless, contacts between Christian Europe and the Islamic world yielded little effect on learning until the second millennium AD.

reflected in the Church hierarchy, but functioned independently and was subject to a different system of law. This parallelism contained the seeds of Church-versus-State conflict, not least because the ecclesiastical lords became major landholders, and land and warfare formed the basis of wealth in feudal society.

Feudal organisation differed in detail from country to country, but it had the following general characteristics:

- The main social division was between the landed and the landless.
- Social action (decisions about events in individual lives) were prescribed rather than chosen – you did as you were told.
- The bulk of society (the peasantry) was more or less identical in status and lifestyle.
- Social change was slow, feudalism tending to retard the development and acceptance of new ideas.

The Power of the Church

After about AD 1000 the waves of barbarian invasions in Europe subsided and some feudal states began to achieve stability. Christian teaching and the enforcement of law played a major part in this stabilisation. Concomitantly, the power of the Church increased. In the middle of the 11th century there was a lasting rift between the Catholic and Orthodox Churches. The former, centred in Rome, took centralised control of religion throughout western and northern Europe. The latter, centred in Constantinople, controlled religion in Greece and much of eastern Europe.

'Control of religion' was vastly important. The Church regulated all aspects of life in the present world and prepared souls for the next. Its political, economic and psychological power became all-pervading. Kings ruled by the Church's consent, a king's first duty – before the duty of military defence of his kingdom – being to uphold the authority of the Church, giving us the real meaning of the phrase 'Divine Right of Kings'. The feudal organisation of society enabled the Church to gain this monolithic power.

The Islamic golden age proved that a unifying monotheism can be compatible with freedom of thought and innovative exploration of the natural world. The same held in Christian Europe; the idea that the Church suppressed or opposed all inquiry is largely false. Nevertheless, as in the Muslim world, thinking was dominated by religion: Christian theology was the basis of all understanding. Nowadays, we would reject any claim about the natural world that contravened the laws of physics, as judged by leading scientists. Nine centuries ago we would have rejected any claim about the natural world that contravened the Bible and the writings of the Church Fathers, as interpreted by ecclesiastical authority.

The most venerated and influential of the Church Fathers was Augustine of Hippo (354–430). A Neoplatonist in his youth, Augustine was converted to Christianity. Although his teachings retained a tinge of Neoplatonism, the conversion made him intolerant of anything that deviated from Christian orthodoxy as he perceived it. He attacked all other interpretations of Christianity (Manicheism, Donatism, Pelagianism etc.) as 'heresies', using his own slant on the Scriptures to support his case. His criticisms were so effective that later theologians considered his authority sufficient for deciding most doctrinal issues.

In the *Confessions*, primarily a devotional work, Augustine grappled with problems such as the origin of Creation, and the nature and measurement of time. He interpreted Plato's Forms as eternal ideas in the mind of God and inferred that everything that exists in space and time is imperfect. However, he regarded the natural world as a reservoir of parables for our instruction. Humans, as God's special creation, were at the centre of the cosmos and everything in the world was created for our instruction as well as our practical use and benefit. Nature was to be observed and studied carefully, but the aim was not objective description; it was to find moral lessons by which our lives could be guided and improved. Thanks largely to Augustine, mediaeval accounts of animate nature combine precise and detailed observation of reality with accounts of wholly imaginary animals, all to support moral homilies. They are entirely different from anything we could call 'scientific'.

The Universities and Scholasticism

Little more than a century after the barbarian invasions of Europe had ceased, universities were founded in Europe. The earliest were Bologna, Padua, Montpellier and Paris; these pioneering institutions rapidly developed offshoots, so that by 1200 almost every western European country had at least one university. The purpose of these institutions was to train people in the professions: theology, law and medicine. Basic education was in the seven liberal arts defined by Isidore and Alcuin. Roman and ecclesiastical law, Galenic medicine and the works of the Church Fathers were standard texts, along with the Scriptures.

The universities were corporate bodies that had certain immunities in law. The Church recognised that students needed to discuss knowledge and beliefs before they could achieve true (acceptable) understanding. Therefore, it was possible in a university to debate ideas, even heretical ones. This structure fostered a hunger for fresh learning, for which there was a ready source: the Arabic translations of, and commentaries on, classical philosophy, especially those housed in Córdoba in Islamic Spain. Translation of Arabic texts into Latin became an industry in the late 12th and 13th centuries. Foremost among the texts were those of Averroës and Maimonides, though by 1200 there were reasonably accurate Latin translations of major works by Aristotle, Plato, Euclid, Ptolemy and Galen, and fragments of other classical authors such as Archimedes. Later in the 13th century, more reliable (Greek) versions of classical works housed in Constantinople, especially those by Aristotle, were translated into Latin.

The universities established a particular style of learning, referred to as *Scholasticism*. The Scholastic approach was one of rational debate or disputation. Its aim was to present doctrine clearly, concisely and accurately. Scholastic debate had the following general structure: first propose a question, then state various arguments pointing to a solution opposite to your own, then give your own solution, and finally refute the contrary arguments. A Bachelor's degree candidate 'determined', that is, put forward propositions and defended them against opponents. Two or three years' study, and more 'determining', converted him into a Licentiate. Further delay and further disputation saw him at last a Master. A few Masters became Doctors of the Church. The propositions chosen for debate concerned topics of central importance in scholasticism; for example, the Aristotelian distinction between *matter* and *form*, and the question of universals.

The question of universals hinged on a quotation from the Neoplatonist Porphyry, which fired enthusiasm in Europe's universities[11]:

> Now concerning genera and species, whether they be substances or mere concepts of the mind; and if substances, whether they be corporeal or incorporeal, and whether they exist apart from sensible things or in and about sensible things, all this I will decline to say.

Some thinkers (*realists*), following Plato, believed that universals were real; others, *conceptualists* such as Peter Abelard (1079–1142), held that they existed only as mental constructs; still others, *nominalists* such as Roscelin (1050–1125), considered them to be no more than words. Each of these views found support in the centuries to follow, and each has supporters today.

Scholastic thought reflected the essentially static order of feudal society. Change was not ignored, but the emphasis was upon what was *established* as a goal to be sought (e.g. perfect morality) or an achievement already realized (e.g. a full-grown society), not the development or evolution of either. The relationship between the world and God was, of course, a major issue. The general consensus was that God is 'high above all nations'. The world has its being, but this is infinitely inferior to the Being of God. The world owes its origin to a free volition of God, put forth at the beginning of time, created out of nothing; and it owes its continued existence to the good pleasure of its Creator.

Regarding knowledge of the observable world, the Scholastic thinkers were predominantly *empiricists*; that is, they held that all knowledge of the natural world begins from what the senses perceive. This made them receptive to the works of the Islamic scholars, increasingly available as Latin translations, and of course to those of Classical antiquity.

Pro and Contra Aristotle

The impact of Averroës (and therefore Aristotle) on Christian thought was dramatic but complex. On the one hand, translated into Latin, the Averroist idea that faith and reason were compatible fell on prepared ground. Anselm of Canterbury (1033–1109), the 'Father of Scholasticism', with his outstanding philosophical and theological talents,[12] greatly influenced the style of thought in the universities. Following

---

[11] Guthrie K (1989) Porphyry's Launching-Points to the Realm of Mind: an Introduction to the Neoplatonic Philosophy of Plotinus. Phanes Press, York Beach ME.

[12] Anselm was the author of the 'ontological argument for the existence of God', which has divided philosophical opinion ever since and is still much debated.

Averroës, he insisted on the importance of reason to the life of faith – beliefs need to be rational. Averroës also offered a critique of Neoplatonism that attracted many Christian thinkers. More generally, Aristotle's logic, his account of the structure of the universe and his discussions of politics and law won many adherents in 13th century Europe, a group who later became known as the 'Latin Averroists'. A prominent example was Siger of Brabant (*c*.1240–*c*.1285) at the University of Paris.

On the other hand, Averroës had made claims that seemed incompatible with Christian teaching, three of which were salient. First, he claimed that Aristotle's philosophy denied personal immortality, contradicting Christian teaching on the afterlife. Second, he asserted that the world was eternal, apparently denying the Creation. Third, he held that all humanity shares one soul, in which individuals participate. This contradicted Christian teaching about the soul, with its individual faculties of mind and will. For these reasons, several scholars denied that Aristotle's teachings could be harmonised with Christian theology. Bonaventure (1221–1274) was prominent here; he used elements of Aristotle's thought in his arguments, but he was essentially a follower of Augustine and a Platonist at heart. Others were more extreme, such as the anti-Aristotelian scholar Peter John Olivi (1248–1298).

Throughout the 13th century, the works of Averroës and Aristotle were repeatedly banned by the Church authorities, and as repeatedly re-legitimised. It was a time of intense intellectual ferment and original thought throughout Catholic Europe, and everyone recognised the debt that the new universities owed to the achievements of the Islamic golden age. In this climate, some foundations were laid for what in centuries to come would be recognised as modern science.

Albertus Magnus and Thomas Aquinas

An outstanding, encyclopaedic thinker of the newly formed Dominican order, Albertus Magnus (1193–1280) believed that philosophy and religion could and should co-exist peacefully. He played a major role in introducing Greek and Islamic thought into the universities, though he was uncomfortable about some Aristotelian theses. He famously remarked that

> 'Science does not consist in ratifying what others say, but of searching for the causes of phenomena'.[13]

Here, we should understand 'science' as denoting knowledge in general. For Albertus and other Scholastics, the principal and fundamental science was theology. He also had consuming interests in astrology, alchemy and magic.

Thomas Aquinas (*c*.1225–*c*.1274), the most outstanding pupil of Albertus, showed prodigious intellectual prowess even before he entered the University of

---

[13] Quoted in Walsh J (1907) The Thirteenth, Greatest of Centuries. Catholic Summer School Press, New York, Chapter 3.

Naples at the age of 11 to study the liberal arts. He became a Dominican at the age of 15 and four or five years later was placed under the supervision of Albertus. He visited Paris with Albertus and there he ultimately became a master of theology.

Ranked along with Augustine as an exponent of Christianity, Aquinas is important for our inquiry in two main respects. First, he dissected the 'true Aristotle' from the 'errors' of Averroës and made mediaeval Church teaching firmly Aristotelian rather than Platonist. Second, he developed the Neoplatonist notion of the great chain of being into a pyramid with God at the apex and inanimate nature at the base (Fig. 4.1). This concept exactly reflected the structure of feudal society, with the king at the apex and the peasants at the base, mirroring the situation of the

Fig. 4.1 The great chain of being, the Christianised Neoplatonist hierarchy relating the highest to the lowest in the cosmos. God is seated above the various ranks of angels. Below the angels are humans (exclusively male), then come the higher and lower animals, then the plants, then the inanimate world of earth and rocks. This is the most famous visual representation of the great chain of being, engraved by the Franciscan Didacus Valades in his *Rhetorica Christiana* (1579) after his return from missionary work in Mexico

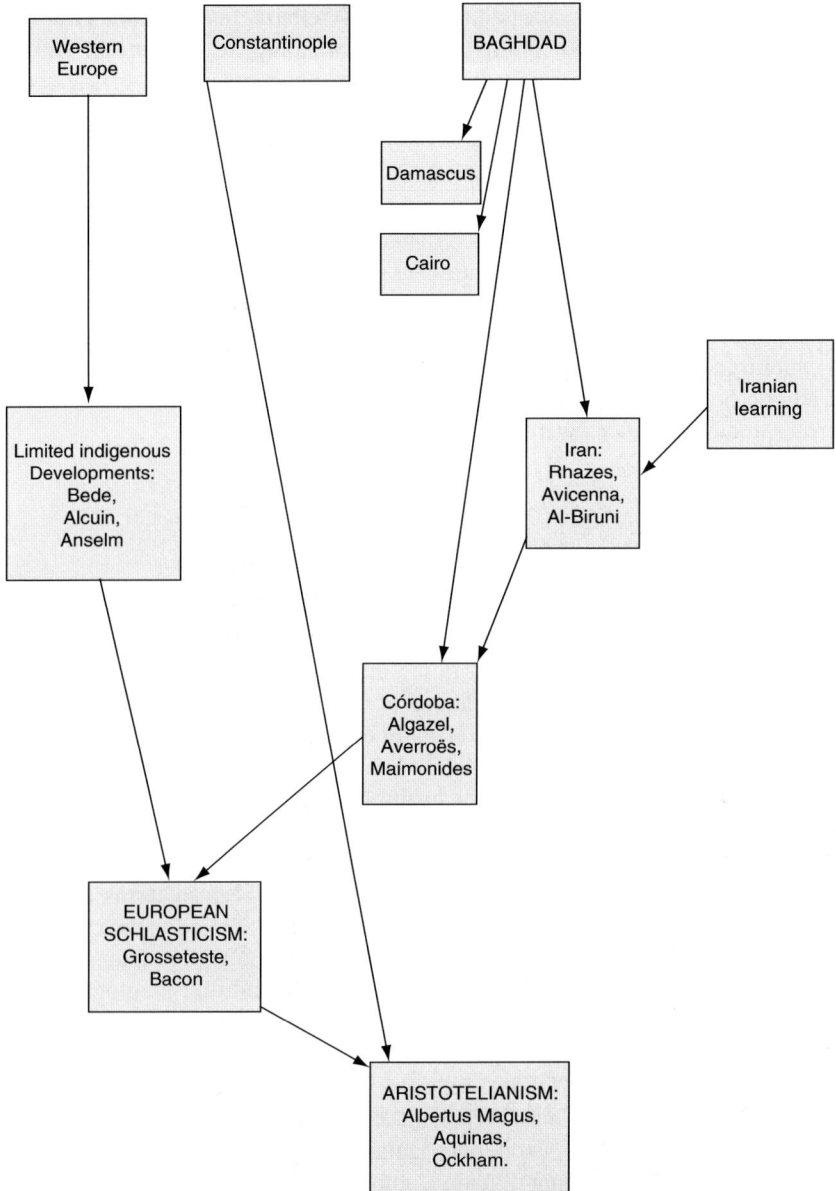

Fig. 4.2 Summary of the routes via which Classical learning was fostered and advanced during the Islamic golden age, preserved in Constantinople and assimilated into the Scholastic tradition of the European universities during the 12th and succeeding centuries

Church, with the Pope at the apex and the laity at the base. In effect, this gave feudal structures the appearance of philosophical and theological legitimacy. His great *Summa Theologica* was unfinished at his death, but remains one of the seminal works of Christian theology.

Well versed in the Islamic writings, Aquinas wrote a treatise on alchemy. Perhaps his most salient contribution to natural philosophy (as opposed to theological philosophy, his main concern) was an extension of the Averroist notion of inertial mass. Averroës had introduced this concept in relation only to celestial bodies, but Aquinas applied it to all material bodies, a significant step towards the science of mechanics.

Although many Aristotelian propositions were again banned in Paris three years after Aquinas died, the ban was short-lived. A generation later, Thomas Aquinas was canonised, and Aristotelianism was unassailably established as an integral part of Catholic belief. The debates of 13th century Europe about the acceptability of Averroës and Maimonides were finally over.

Aristotelianism

One reason why Augustine and other Christian Fathers were accepted more or less uncritically was that they had lived significantly nearer to the time of the Garden of Eden than their latter-day successors. Being therefore closer to perfection, they had had clearer understanding of the world and of God's purposes. The same applied to Plato and Aristotle, even though they had not enjoyed the benefit of Christian revelation.[14] Once it was accepted that Aristotle's writings cohered with Christian theology, his acceptance was assured. For example, Aristotle's spherical Earth at the centre of a cosmos of concentric, interacting spheres placed humanity at the centre of creation. The 'prime mover' beyond the sphere of the fixed stars was identifiable with God. The Creator and His work could readily be seen in terms of the four Aristotelian causes (material, formal, efficient and final) applied to the cosmos as a whole and to each of its parts. In particular, details of the Divine Plan lay in final causes, and these were therefore a proper subject of study by theologians. More generally, a model of the cosmos that was comprehensive and complete, essentially static and teleological, was perfectly consistent with the Church's view of God's creation – and, indeed, with feudal culture as a whole.

It is small wonder that Church teaching and Aristotelianism became so closely entwined that to challenge Aristotle was tantamount to challenging the authority of the Church. A complete account of nature, admitting no fundamental challenges, was the perfect complement to the Scriptures.[15] Scholastic thought was not 'naturalistic',

---

[14] John of Salisbury (*c*.1115–*c*.1180), one of the first European thinkers to be explicitly influenced by Aristotle, asserted that the Ancients – albeit pagan – had clearer understanding than men of latter years because they were so much closer to the time of human perfection, before the Fall of Man.

[15] In the Statues of Oxford in the 14th century we find: '*Bachelors and Masters of Arts who do not follow Aristotle's philosophy are subject to a fine of 5 shillings for each point of divergence.*'

but the thinkers of the middle ages continued the Islamic tradition of learning and actively extended it, making contributions to what would later become 'science'.

One non-naturalistic feature of mediaeval thought was an *organic* or 'animistic' view of the natural world. Alchemy, for instance, was primarily about the 'spirits' that distinguished one substance from another. The transmutation of metals would require the replacement of one spirit by another. (Our language today retains traces of this 'organic' perspective – for example, we commonly speak about 'Mother Earth' and 'veins of ore'.) By the same token, the mediaeval viewpoint was *teleological* rather than mechanistic; like Aristotle before them, the Scholastic thinkers ascribed purpose to events in the natural world.

Robert Grosseteste and Roger Bacon

These two pioneers of the English intellectual tradition at Oxford were peripheral to the debates about Aristotle and Averroës raging in Paris, but were instrumental in picking up the threads of learning from the Islamic world and introducing them to European scholarship. Grosseteste (*c*.1175–*c*.1253) is now best known for his ecclesiastical and political endeavours, defending the interests of the Church against the encroachments of the State, and for his discussions of Biblical texts. However, his contemporaries and immediate successors saw him as the pioneer of a new intellectual movement, a versatile thinker who expounded Muslim discoveries in mathematics and physics and stimulated interest in the study of the natural world for its own sake. His many treatises on natural philosophy included studies of light (including rainbows), aspects of astronomy, the tides, and mathematical reasoning.

Grosseteste's reputation has been somewhat eclipsed by that of his illustrious pupil, Roger Bacon (*c*.1214–*c*.1292). Bacon was a wide-ranging if unconventional thinker who made no secret of his debt to the Islamic golden age. In his *Opus Majus* (1266), which argued that the philosophy of Aristotle and Grosseteste's 'new' way of thought could be incorporated into theology, Bacon wrote:

> Neglect of mathematics works injury to all knowledge, since he who is ignorant of it cannot know the other sciences or the things of this world.

He saw practical ideas arising from his intellectual interests: the design of spectacles from optics and the refraction of light, calendar reform from the study of astronomy. He was enthusiastic about Islamic alchemy, including its 'magical', Hermetic roots, and was a pioneer of European alchemy. He emphasised the importance of experiment (including precise observation of nature), a radical notion[16]:

> Reason... reaches a conclusion and compels us to admit it, but it neither makes us certain nor annihilates doubt ... Even if a man who has never seen fire proves by good reasoning

---

[16] *On Experimental Science* (1268). We must, however, remember that *experimentum* may be translated as 'observation' or 'experience' rather than 'experiment' in the modern sense. Bacon was voicing Scholastic empiricism rather than, or as well as, echoing Geber's emphasis on the need for 'experiment'.

that fire burns, and devours and destroys things, nevertheless the mind of one hearing his arguments would never be convinced, nor would he avoid fire until he puts his hand or some combustible thing into it in order to prove by experiment what the argument taught... Hence argument is not enough, but experience is.

William of Ockham (c.1282–c.1348)

Although William of Ockham is not remembered as a contributor to 'science' *per se*, he remains an important figure in the philosophy of knowledge. He is often labelled a 'nominalist' but is better described as a 'conceptualist'. He argued that statements containing general terms, e.g. 'men are animals', were convenient summaries of individual statements and of ideas. 'Men' means all men considered as individuals. 'Animals' denotes a mental construct, an idea of 'animality' that can be resolved into characteristics such as life and sensibility, which again are ideas in the mind. Here, Ockham parted company with Aquinas and with realists of all kinds, for whom 'ideas' had a real, Platonic existence. By the same token, Ockham rejected all arguments that purported to prove the existence of God by reason; belief in God depends on faith alone. This debate was to re-emerge in the 18th and 19th centuries as philosophers struggled to understand the relationship between new scientific concepts and the observable world they purported to describe and explain.

Ockham's most famous contribution to the philosophy of knowledge is his statement of the 'principle of simplicity', often known as 'Ockham's Razor': *Pluralitas non ponenda sine necessitate*, which may conveniently be translated *'It is vain to do with more what can be done with less'*. In other words, we should not believe in any entity or consider any hypothesis unless there is compelling reason to do so. Sadly, this 'soundbite' has been bandied around so much that its true import is seldom appreciated.[17]

The Black Death

The most devastating plague ever to hit the Western world wiped out about half the population of Europe between 1347 and 1350. It was one of many outbreaks of this mysterious disease[18] but by far the worst. The resulting catastrophic decline in population had severe economic consequences. Valuable artisan skills became scarce.

---

[17] Bertrand Russell (1946) explicated Ockham's Razor as follows: '... *if everything in some science can be interpreted without assuming this or that hypothetical entity, there is no ground for assuming it. I have myself found this a most fruitful principle in logical analysis*'.

[18] Since the late 19th century it has been conventional to attribute the Black Death to bubonic plague, which is transmitted by rat fleas. In fact, the epidemiology and symptomatology are not consistent with bubonic plague. The disease was more likely to have been a haemorrhagic fever comparable with Ebola.

Poor labourers were able to demand improved wages and conditions. Serfs began to leave their land; crops were not planted or harvested, farm animals strayed and starvation threatened. The lack of social organisation and cohesion promoted lawlessness. The feudal economic, social and political structure based on land-based wealth was irreparably damaged. Portable wealth in the form of money, skills and services emerged. Small towns and cities grew while large estates and manors began to collapse.

The Church suffered considerable injury. It lost not only personnel at all levels, but also prestige, spiritual authority and leadership over the people because it could provide no satisfactory explanation for or prevention of the plague. People prayed to God and begged for forgiveness, but many clergy abandoned their Christian duties and fled. After the plague ended, angry and frustrated villagers started to revolt against the Church.

Intellectual life also suffered, but that setback was temporary, whereas it took many years for the population and the economy to recover. However, the tradition of thought inherited from the Islamic world was actively resumed with scarcely a pause.

The Decline of Islamic Pre-Eminence

Europe had taken on the mantle of learning, and its new intellectual tradition survived the Black Death. There were prominent Muslim contributors to natural philosophy after the 12th century, but the Islamic golden age was over. Among the reasons for that decline we may cite internal conflicts, the devastation wrought by the Crusades, the depredations of Genghis Khan and his successors in the East, and of course the Black Death, the effects of which were by no means confined to Europe.

It is perhaps salutary to reflect that the pioneering age of naturalistic thought in the Greek world lasted from the 6th century BC to perhaps the 1st century AD; that Alexandria remained the centre of intellectual life from around 300 BC to 200 AD; and that the Islamic golden age had also survived for about half a millennium, from the 8th to the 13th centuries. Modern science is often considered to have originated in the mid-16th century, so it has been at the forefront of our intellectual life for some 450 years.

Is its life expectancy likely to differ from that of its predecessors?

Chapter 5
The Scientific Revolution

Traditionally, the 'scientific worldview' is said to have originated in Western Europe in 1543, when Copernicus's *On the Revolutions of the Heavenly Bodies* was published in Vienna – a book that not only introduced much of literate Europe to the notion of a 'heliocentric universe', but was also contrary to Aristotelian teaching. In 1686, Newton published his *Principia*, which put a final nail in the coffin of Aristotle's physics. The 143-year period between these two publications is usually known as the *Scientific Revolution*.

The label 'Scientific Revolution' is not meaningless – the 16th and 17th centuries indeed witnessed a dramatic intellectual transition, as we indicated in Chapter 1 – but for the following reasons it can mislead:

- Copernicus and his successors inherited the tradition of Classical learning that had continued, transmuted but unbroken, through the Islamic golden age and late mediaeval Europe (see Chapter 4 and below).
- European views of the natural world had not become entirely 'modern' by Newton's time.
- Most importantly, the process of change was underpinned by a wider cultural transformation.

The Continuing Scholastic Tradition

Buridan

Jean Buridan[1] (1300–1358) studied in Paris under William of Ockham. Even a slight acquaintance with Buridan's work is enough to disprove the idea that the new way of thinking about the natural world sprang into being, unheralded, in 1543.

---

[1] One of the most highly regarded philosophers of his age, Buridan is now remembered – if at all – for the thought experiment known as 'Buridan's Ass', a very early gesture towards the theory of probability: put an ass exactly equidistant between two bales of hay and it will starve to death because it will be unable to decide which one to eat.

P.S. Agutter, D.N. Wheatley, *Thinking about Life*,
© Springer Science+Business Media B.V. 2008

Buridan developed the concept of *impetus*, which led immediately to the notions of *inertial mass* and *momentum* familiar in Newtonian mechanics. He said that 'impetus' increases with the speed with which an object is travelling and with the amount of matter it contains. Moreover, a moving body can be arrested only by opposing forces such as air resistance and gravity:

> ...after leaving the arm of the thrower, the projectile would be moved by an impetus given to it by the thrower and would continue to be moved... were it not diminished and corrupted by a contrary force resisting it or by something inclining it to a contrary motion.[2]

This claim was alien to Aristotelian physics, which maintained that a body remains in motion only when a continuous external force is applied. 'Impetus' was therefore a development of the idea of 'force' explored by Averroës. It also echoed the teachings of the 6th century writer John Philoponus.[3] But Buridan saw it as a *correction* of Aristotle, not a challenge to Aristotelianism; he still maintained that there is a fundamental difference between motion and rest. He also believed, like a true Aristotelian, that impetus gives rise to linear motion in the terrestrial part of the cosmos but to circular motion in the celestial part.

Oresme

The polymath Nicole Oresme (*c*.1323–*c*.1382) – economist, mathematician, physicist, astronomer, philosopher, psychologist, musicologist, theologian, bishop, translator, counsellor of Charles V of France – was a major contributor to the development of mathematical and mechanistic thought, though still within the Aristotelian tradition. Highly critical of astrology, he grasped the notion of 'law of nature' and distinguished the necessary from the contingent. He pioneered the use of 'Cartesian coordinates' (graphs) and related them to the equation for a straight line, following the precedent of Thabit bin Qurrah (Chapter 4) and anticipating the analytical geometry of Descartes by almost three centuries. Likewise, he adopted and developed al-Biruni's concept of *acceleration*, which is usually attributed to Galileo.[4]

Using his innovations in music theory, Oresme developed a method of calculating with fractional irrational exponents, a remarkable advance in arithmetic. (We recall that the quadrivium portion of the liberal arts linked music with arithmetic

---

[2] Quoted in Thijssen JMMH, Zupko J (eds) (2001) The Metaphysics and Natural Philosophy of John Buridan. Brill, Leiden.

[3] Philoponus (*c*.590–*c*.650) disproved Aristotle's belief that the speed of a moving body is proportional to its weight and indirectly proportional to the density of the medium, using the same experiment that Galileo allegedly performed centuries later. His work was placed under anathema by the Church for 600 years but was rediscovered in the 1200s by Bonaventure and others; it was presumably known to William of Ockham and thus to Buridan.

[4] More than a generation before Galileo, Domingo de Soto (1494–1560) applied Oresme's law to the uniformly accelerated falling of heavy bodies and to the uniformly decreasing ascension of projectiles.

and geometry.) He also developed new ideas about limits, threshold values and infinite series, anticipating the infinitesimal calculus of Newton and Leibnitz.

In much of his physics, he accepted and transcended the arguments of Buridan. In opposition to Aristotle, who said that heavy bodies are naturally located at the centre of the world and light bodies in the concavity of the moon's orbit, Oresme proposed that

> The elements tend to dispose themselves in such a manner that, from the centre to the periphery, their specific weight diminishes by degrees.[5]

This was exactly the argument later advanced against Aristotle by followers of Copernicus, particularly Giordano Bruno (Bruno was clearly familiar with Oresme's work, though he never acknowledged the precedent). Even more strikingly 'Copernican' was Oresme's discussion of the rotation of the Earth, a reworking of al-Biruni's argument (Chapter 4). No experiment, he wrote, can decide whether the heavens move from east to west or the Earth from west to east – the senses can never establish more than one relative motion. He showed that Aristotle's arguments against the movement of the earth were invalid, then addressed objections based on Biblical texts, articulating rules of interpretation that are still followed in Catholic exegesis today. Finally, he invoked Ockham's Razor to support the claim that the earth moves rather than the heavens. Oresme's style of argument is typically Scholastic (Chapter 4), but it shows that even in an intellectual climate dominated by Artistotelianism, it had become possible to criticise The Philosopher.

More remarkable still was Oresme's work on optics. He showed that the atmosphere refracts light, and more generally that the amount of refraction increases with the density of the medium, a fact that had not only escaped both Alhazen and Witelo, but was later to escape Kepler. He also showed, anticipating Newton, that white light is an amalgam of the colours of the spectrum. Once again his knowledge of music theory provided a metaphor – light is a mixture of colours just as sound is a mixture of overtones.

Perhaps the most striking feature of Oresme's contributions to natural philosophy was his willingness to look for causes in the natural world itself rather than invoking God – or any intrinsic animism. This heralded the *mechanistic* philosophy of Galileo and Descartes. It was perhaps the first deep philosophical departure from mediaeval Aristotelianism – yet in spirit, true to Aristotle.

Nicholas of Cusa

A polymath of the following century, Nicholas of Cusa (1401–1464) came to be much more widely read than Buridan or Oresme; some of his works remained in

---

[5] *Traité du ciel et du monde* (1377), written at the request of Charles V. Oresme has been dubbed 'the Einstein of the 14th century'. To describe Einstein as 'the Oresme of the 20th century' would convey little or nothing to most of us – which illustrates how little we know and appreciate our intellectual heritage!

print for over 100 years. Copernicus, Bruno, Kepler and Galileo all knew his
writings. Such was the impact of the printing press on scholarship.[6]

Contrary to the Aristotelian belief that celestial orbits are perfectly circular,
Nicholas said that no perfect circle can exist in the universe. It is little wonder that
Kepler, who discovered by observation that the planetary orbits are elliptical, called
him 'divinely inspired'. Perhaps more radically, Nicholas denied that the universe
is finite or that the Earth occupies a central position in it, assertions for which
Bruno would be executed for heresy in 1600. Nicholas also developed Oresme's
concept of infinitesimals (paving the way for the calculus) and his predecessors'
work in optics; he may have been the first since Al-Haytham to use concave lenses
to correct myopia.

The Mertonian Calculators

A group of fellows of Merton College, Oxford during the early and mid-14th century
(Thomas Bradwardine, John Dumbleton, William Heytesbury, Richard Kilvington
and Richard Swineshead) made advances in pure mathematics that were later to
become relevant to mechanics. Their points of departure were Aristotelian logic
and physics, but they were intent on quantifying every physical observable, including heat and force, in contrast to Aristotle, who had believed that only lengths and
motion could be quantified. They developed the pioneering work of Al-Battani on
trigonometry (Chapter 4). Most famously, they (notably Heytesbury) proved the
mean speed theorem, later to be attributed to Galileo.[7] The context of their work,
like that of Buridan, Oresme and Nicholas of Cusa, was Aristotelian scholasticism,
and their interests lay more in philosophy and logic than in the natural world *per se*.
For example, one of their followers, Walter Burley, used their arguments to develop
an extreme realist position – in deliberate opposition to William of Ockham –
holding that unless the abstract or general entities that were the subjects of disputation had real existence, then no consistent knowledge of the external world could
be obtained.

The work of the Calculators, like that of Buridan, proves that the Black Death
did not seriously disrupt the intellectual tradition that we have surveyed. Several of
the Calculators were victims of the plague (as was William of Ockham), but the
learning that they fostered survived. The wider culture, however, was to undergo
permanent change.

---

[6] The printing press enabled knowledge to spread rapidly (it cost less to make and distribute copies
of books, and literacy increased) and, more significantly, ensured that unlimited numbers of copies
of any one book were *identical*.

[7] If an object travels from A to B at uniform acceleration, the time it takes to reach B is the same
as the time it would have taken had it travelled all the way at its average (constant) speed.

Wider Cultural Changes: Secularisation

After the Black Death, feudal culture declined and 'early capitalism' began to emerge in Western European countries. This profound cultural shift, known as *secularisation* or *rationalisation*, was intensified and probably accelerated by the discovery of the Americas,[8] the Protestant Reformation,[9] and a number of technological innovations of which the printing press had the most dramatic and far-reaching effects. The main impact of secularisation on Europe occurred between 1450 and 1600, but its beginnings (in so far as such broad processes have 'beginnings') lay in the 14th century.

Secularisation was multifaceted, but it can be summarised as follows:

- The main social division changed from landed versus landless, as it had been in feudal society, to employer versus employed. Concomitantly, wealth came to depend on manufacture and trade rather than land ownership and warfare. The thinkers of the Scientific Revolution lived in a world increasingly dominated by machinery and clocks, not by the works of the Creator. The achievements of an individual were seen increasingly as the results of that individual's efforts. The *Protestant Work Ethic* was born.
- Social differentiation increased. In mediaeval Europe the vast bulk of the population were agricultural labourers – slaves, more or less. As manufacture and trade increased and towns and cities grew more numerous and populous, more kinds of employment and lifestyle became available. Most people were still agricultural labourers, but there was a steady increase in the variety of career opportunities.
- Concomitantly, social action became less a matter of prescription and more a matter of choice. Almost all mediaeval people lived where they were born, married whom they were told to marry and worked as ordained by the feudal overlord. In the early capitalist era they could move to town and enjoy some freedom of choice in respect of home, spouse and job. A popular saying in the 15th and 16th centuries was 'The air of the town is the air of freedom'. This aspect of secularisation is sometimes called *individuation*; it encouraged individual choice. In turn, it may have encouraged the growth of individual ideas about how the world worked.
- Government of the new towns and cities passed increasingly from hereditary feudal land-owners to leading manufacturers and traders. The merchants were intensely competitive; town competed with town, and competition among nations became dominated by trade rather than land acquisition. Nevertheless,

---

[8] This opened contact with wholly new cultures (and great sources of wealth).

[9] The merchant leaders of the towns and cities cherished autonomy and freedom of action. Clashes were inevitable, not only between the wealthy merchants of German and Flemish cities but also with the Church's tax-gatherers. This was a major factor in the Protestant Reformation. Importantly, it occurred in the west, not the east, of Europe – a reaction against the Catholic not the Orthodox Church.

within their own towns, these successful capitalists made collective political decisions so that no individual or trade was seriously disadvantaged. The idea of modern democracy may have emanated from this practice; certainly it encouraged decision-making by consensus rather than by edict.

- The effective leaders of the new culture, the merchants, became increasingly literate and numerate; that was a matter of survival. Thus, literacy was no longer confined to the Church and the university-educated.
- The power of the Church waned while that of the state increased. Because the mediaeval Church had exerted such monolithic power, the decline of its influence made changes of world-view not only possible but well-nigh inevitable.
- Change became institutionalised. In feudal Europe, one year was much like another, every generation like the preceding one. The social order was essentially static; that is not to say that it never changed, but rather that change – progress – was resisted, not encouraged, by the socio-economic structure. Capitalist culture, in contrast, depended on innovation and the creation and exploitation of markets – in a word, change. Capitalist societies are *intrinsically progressive*, which coincidentally is one of the hallmarks of science (Chapter 1).

The Renaissance

Early in the secularisation process, a corollary movement arose (initially in Italy, particularly Florence) that was to have a marked effect on European thought. Nineteenth century historians dubbed this movement 'the Renaissance'. Those who were active in it during the 15th and early 16th centuries called themselves *umanisti*, because their focus was on *people* rather than the Church. The pronouncement[10] that 'Man is the measure of all things' stood as their watchword.

- The *umanisti* were first and foremost merchants. They are remembered for their seminal contributions to the arts, but their interests were primarily practical – the marketplace and the potential of new technologies.
- They consciously dissociated themselves from the Scholastic tradition of the universities and therefore from Aristotelianism while remaining staunchly Christian.
- In the quest for a Classical basis for their values and attitudes, they returned to Roman writings and to Roman technology (particularly civil engineering). The intensity with which they espoused their assumed Roman roots is reflected in the personal letters they wrote to Cicero and other luminaries of the Roman era. (They 'discovered' the ancient Greek classics later in the Renaissance period.)

---

[10] This aphorism was due originally to Protagoras of Thrace (*c*.480–*c*.411 BC), often regarded as the first of the sophists. Protagoras believed that there is no objective truth or falsity but that each individual is his own authority. The *umanisti* interpreted the 'slogan' more broadly; thus, the famous Renaissance buildings of Florence were built to a human scale, in contrast to the Gothic churches of the period, which were deliberately built to a superhuman scale.

- In consequence, they tended to discount not only most contemporaneous intellectual activity in the universities, but also the entire work of the Islamic golden age. Repudiating their actual inheritance, they constructed an artificial 'intellectual history' that suited their orientation.
- They established an independent tradition of scholarship, founded on their interpretation of Classical antecedents and largely orientated towards practical (military, civil, etc.) as well as artistic concerns.

The Renaissance was an intellectual by-product of secularisation. It contributed in important ways to the emergence of science. In particular, it established the close relationship between knowledge and technology that is peculiar to our modern worldview. It was also responsible for the tacit denial that Muslim thinkers and mediaeval scholars contributed significantly to our understanding of the world. This tacit denial has typified Western histories of ideas since the 17th century.

After the fall of Constantinople in 1453, a number of organisations called 'Academies' appeared, particularly in Italy, devoted to the study and development of ideas found in the newly-available Classical Greek texts. These 'Academies', intended (as their collective title indicates) to emulate Plato's original establishment, were products of the Renaissance: they flourished outside the university sector and independently of the Church. Among the works they studied were texts by Archimedes, which offered mathematical accounts of the physical world. The conviction that mathematics might be more than a convenient descriptive tool took hold and became one of the mainsprings of the Scientific Revolution.

Changing Styles of Thought

Feudalism, already undermined by the social disruption caused by the Black Death, had virtually been eliminated from Western Europe by the 16th century. Progress, individualism and reliance on innovation and change had become socially institutionalised. In the northern part of the continent at least, the power of the Catholic Church was collapsing irreversibly.[11]

Intellectual culture changes with the rest of society. The static, Church-linked world-view of Aristotelianism was becoming more and more anachronistic. Secularisation created conditions that broadly resembled those in the eastern Greek city states around 600–500 BC with their secular merchant power, wide-ranging contacts based on trade, decentralised and non-bureaucratic town governments, competition and literacy. As we have seen, those conditions can foster the emergence

---

[11] The Protestant Reformation brought about a new conception of individual worth and responsibility. Men could claim to be in possession of certain partial truths, including scientific propositions, which Church authorities may not have the competence to judge. Before Luther, such an attitude would have been incomprehensible; but the Scientific Revolution could not have occurred without it.

of naturalistic thought. However, it would be absurd to claim that our modern scientific world-view was directly 'caused' by secularisation.

In fact, *many* alternative new world-views emerged. The Renaissance fostered traditions of magic and mysticism[12] as well as mechanism. Throughout the 16th and 17th centuries these contrasting traditions co-existed, and indeed interacted. Newton's work, to cite a salient example, was strongly influenced by alchemy and mysticism; indeed, Newton wrote much more about alchemy than about mechanics or optics. Part of the reason why the mechanistic tradition, the direct ancestor of modern science, outlasted its rivals may have been its increasing compatibility with secularized, capitalist society. It could accommodate not only the natural world but also a social world of more and more sophisticated machinery. Our scientific world-view may have 'evolved' by something akin to natural selection. It was better adapted to the emerging social and economic environment than any alternative metaphysics of the post-Renaissance centuries.

The Demise of Aristotelianism

Nicolas Copernicus (1473–1543)

In the preface to *On the Rotations of the Heavenly Bodies* was a disclaimer to the effect that the heliocentric model of the universe was not intended to describe reality – it was just a convenient way of doing calculations. Was the disclaimer approved by Copernicus or was it inserted by the publisher? No one knows. What *is* known is that Copernicus had been under pressure from the Vatican to publish his wonderful new ideas[13] for around two decades. Antagonism to Copernicus came initially from the new Protestant leaders, particularly Martin Luther, not from the Cardinals, so his book was published in Roman Catholic Austria, not Protestant Germany. Antagonism from the Vatican came later, when most informed parties came to accept the Copernican account as 'real'.

Tycho Brahé (1546–1601)

As the 16th century wore on, increasing numbers of astronomers in both Catholic and Protestant states began to take the Copernican model seriously. The Danish astronomer

---

[12] For example, the writings of the Hermetic tradition (see Chapter 3) were widely read and they influenced, among others, Bruno.

[13] Not really new, of course – Aristarchos of Samos, Brahmagupta and al-Biruni were among those who had advanced heliocentric models – but new to post-Renaissance Europe.

Tycho Brahé tried to come to terms with some of its implications. For instance, if the universe is not geocentric, i.e. if the Earth is not the centre of the cosmos, then the fact that heavy objects fall downwards becomes problematical. It was no longer possible to say, like Aristotle, that objects such as stones in which the element earth predominated had a 'natural tendency' to move towards the centre of the universe, i.e. the centre of the Earth. A new explanation for the phenomenon of gravity[14] was needed.

The problem of gravity was not the only challenge to Aristotle. Brahé, the last great astronomer before the invention of the telescope, discovered a new star, *stella nova*. His treatise on this observation added the word *nova* to our vocabulary. More immediately, it challenged Aristotle's assertion that no change (other than circular motion) could occur in the celestial part of the cosmos. If nothing can be created and nothing can change its properties (such as the ability to shine visibly), how could the *stella nova* be explained?

Equally serious for the Aristotelian account of nature were Brahé's observations on comets. It had been known since very ancient times that cometary orbits were highly elliptical. In other words, they were far from being perfect circles. Aristotle had said that because the motions of comets were imperfect, they belonged to the terrestrial part of the cosmos. Brahé showed convincingly that the comets he studied were more distant from the Earth than the Moon. Therefore, they lay beyond the sphere of the Moon, belonging to the celestial part of the cosmos. Hence, objects in the celestial world could move imperfectly, just as Nicholas of Cusa had reasoned.

Johannes Kepler (1571–1630)

Worse was to follow. Brahé's successor at the imperial court in Prague, Johannes Kepler, went on to develop a wholly new conception of what we now call the solar system. When Kepler went to work with Brahé in 1600 he began to study the orbit of Mars in detail. Brahé died a year later but Kepler retained his post and continued his studies. To his consternation,[15] he discovered that the Martian orbit was not per-fectly circular but elliptical. Within a decade, he had shown that the same was true of *all* the planetary orbits, and thus imperfection in the celestial world was not con-fined to comets. Surprisingly, the intellectual revolutionary *par excellence*, Galileo, ignored Kepler's discovery. To the end of his long life Galileo believed, like Copernicus – and like Aristotle – that the planetary orbits were perfect circles.

---

[14] The term *gravity* was Aristotle's. It meant, simply, the tendency to fall towards the centre of the cosmos. Objects in which the elements air or fire predominated had a natural tendency to move towards the higher terrestrial spheres, a tendency called *levity*.

[15] Kepler had suffered the most insecure of childhoods in a war-torn region of Germany. His earli-est attempts to characterise the structure of the solar system betrayed a deep desire to find perfect order in the natural world. Later in life he delighted in discovering beautiful mathematical regu-larities, including the structures of snowflakes and the phenomenon of phyllotaxis. Finding that the planets were 'imperfect' in their orbits must have distressed him deeply, and his acceptance of the observation testifies to his remarkable intellectual honesty.

Giordano Bruno (1548–1600) and Galileo Galilei (1564–1642)

Bruno was one of the earliest writers to confront Aristotle's teaching, and therefore the Church, with a mathematical account of physical processes. Bruno contrived to be excommunicated by both the Calvinists and the Lutherans and was finally, after seven years of questioning by the Inquisition, burned at the stake. Galileo was initially more circumspect. In his earliest writings (re-workings of Archimedes, in effect), he shared Bruno's commitment to mathematical explanation, but he tried to avoid confrontation. He had personal friends among the Cardinals; one of them later became Pope Urban V. Much of his *Letter to the Grand Duchess Christina* of 1612 is an attempt to reconcile the new mathematical philosophy with the Scriptures. Nevertheless his certainty that mathematics provided the key to understanding the physical universe was unshakeable. '*The Book of Nature,*' he wrote, '*is written in mathematical characters.*'

Galileo transformed the telescope from a toy or a lens-grinder's master-work into a powerful astronomical instrument. Turning his improved telescope on the moon, he saw not a perfect crystalline sphere, as Aristotelian teaching predicted, but a rough surface pitted with craters. No one was willing to believe him. A rough surface, said Galileo's critics, could not reflect light – yet the moon shines brightly. Galileo countered by reasoning that the greater the surface area, the more light is reflected; the rougher the surface, the greater the area; so the rougher the surface, the more light is reflected.[16] This vignette shows how Galileo relied on both observational evidence and reasoning. His message was that we should believe what our senses tell us, even when the data go against our preconceptions, and then explain the data using logic and mathematics.

The telescope also revealed the four largest moons of Jupiter, appearing and disappearing periodically as they orbited the planet. It followed that other planets could have moons, just as the Earth has. This was so at odds with the Aristotelian beliefs of the Church that the Vatican's astronomers, looking through the same telescope, were unable to see Jupiter's moons. Deeply held preconceptions can influence what we see and what we fail to see.

The Principle of Inertia

Galileo and René Descartes (1594–1650) independently formulated the *principle of inertia*, which is often regarded as the death-blow to Aristotelianism.[17] In place of Aristotle's pronouncement 'everything that moves is moved by something', this principle holds that all moving objects continue to move at the same velocity unless

---

[16] We are often unaware of this because the light striking a very rough surface is scattered in all directions and therefore the surface *looks* dull. The silvering of a mirror is extremely rough on a microscopic scale and therefore presents a very high surface area to reflect light.

[17] See, for example, Monod (1970); full reference in bibliography.

something stops them. Of course, Buridan and Oresme had said the same thing two or three centuries earlier, but it was typical of post-Renaissance thought to deny mediaeval (as well as Islamic) precedents. Galileo believed that unimpeded movement was along a circular path, not along a straight line; otherwise, the principle is identical with Newton's First Law of Motion. The principle of inertia explains movement in terms of antecedent mechanical causes that are susceptible to mathematical analysis. Buridan had seen it only as a 'correction' to Aristotle, but in the hands of Galileo and Descartes it broke the traditional reliance on Aristotle's qualitative 'four causes'. In particular, it severed physics from any reference to final causes. Physics became mechanistic, non-teleological, and reliant on mathematics for its explanations. By implication, it removed the need to invoke God in explaining natural phenomena.

Pierre Gassendi (1592–1655)

Gassendi offered a different challenge: he revived Classical Epicurean teaching and with it a belief in Greek atomism. This was an implicit attack on Aristotelianism; Aristotle, we recall, had reasoned that atomism implied the existence of vacuum, and he denied the possibility of vacuum. Gassendi's work was therefore highly controversial. However, when Galileo's private secretary, Evangelisto Torricelli, demonstrated by experiment that a vacuum could be produced above a column of liquid in a sealed tube, and explained the phenomenon mechanistically, opinion shifted in Gassendi's favour.

It was largely thanks to Gassendi that the predominant 17th century view of the material universe came to be 'atoms in a void', a far cry from the mediaeval, Aristotelian picture that had harmonised so perfectly with Church teaching.

Isaac Newton (1642–1727)

Newton gave the world its most famous and durable scientific theory, classical (Newtonian) mechanics. In the process he invented a mathematical tool, the calculus, which has since found an almost unlimited range of applications.[18] Few of Newton's contemporaries fully understood his *Principia (The Mathematical Principles of Natural Philosophy)* but the work exerted vast influence on the generations that followed. It showed that a single coherent set of ideas rooted in one or

---

[18] Gottfried Leibnitz (1646–1716) invented the calculus independently of Newton and used more convenient symbols for it. It seems likely that Newton made the mathematical breakthrough during the 1660s, Leibnitz some ten years later; but Leibnitz wrote about it before Newton published the *Principia*. An unpleasant precedence dispute followed, which Newton continued to pursue even after the death of his supposed rival.

two postulates (e.g. locations in the unbounded space of the universe are independent of time and of the observer's viewpoint) and based on four simple principles (the three laws of motion and the law of universal gravitation) could account exactly and quantitatively for events ranging from the orbital movements of planets to the descent of apples from trees.

Newton's *Principia* completed the demise of Aristotelianism in physics. Classical mechanics applied equally well to the surface of the Earth and the solar system's planetary orbits. It was applicable on all scales, everywhere in the universe.[19] Hence, there was no distinction between the terrestrial and the celestial. However, it would be wrong to imagine that Newton caused Aristotelianism to vanish overnight. A view of the cosmos that had held sway in Europe for so many centuries and had become integrated into so many aspects of culture could not die quickly or easily. One of Newton's many brilliant contemporaries, the Irish philosopher Robert Boyle (1627–1691), defended Aristotle's 'four causes' analysis even while Newton was compiling the *Principia*; and as we shall argue in subsequent chapters, Aristotelianism persisted in descriptions of living organisms long after it had vanished from physics.

It should not be forgotten that Newton's primary concerns were with Biblical interpretation and alchemy. His *Opticks* contains a long discussion of Noah's Flood and the first appearance of a rainbow on Earth. His conception of 'gravity' had alchemical roots. Also, although (like other post-Renaissance thinkers) he made little reference to his Islamic and mediaeval forebears, his central themes in physics were exactly those of al-Biruni and Oresme: mechanics and optics. Even his most outstanding achievement, the calculus, had precedents in the work of Oresme and Nicholas of Cusa. That is not to deny the immense stature of Newton; it is merely to deny that he was *sui generis* (Fig. 5.1).

The Distinctiveness of Natural Philosophy

The way of thinking about the observable universe that matured in Newton's *Principia*, 'natural philosophy' (later called 'experimental philosophy'), was the immediate forebear of modern science. It was a continuation of the intellectual tradition that had led from Classical Greece through Alexandria, the Islamic golden age and the Aristotelianism of late mediaeval Europe. Its exponents denied (or ignored) these antecedents because they consciously sought a *new* way of knowing; yet without that long and distinguished heritage they would have achieved little. At the same time, its newness was undeniable. It was a product of Europe's secularised,

---

[19] From the time that Newton's theory achieved general acceptance (late 18th century) until the early 20th century, this statement was accepted as true without reservation. For most practical purposes, Newtonian mechanics continues to provide good solutions to physical and engineering problems. Only at the subatomic and ultra-large scales does the theory fail, requiring recourse (respectively) to quantum and relativistic mechanics.

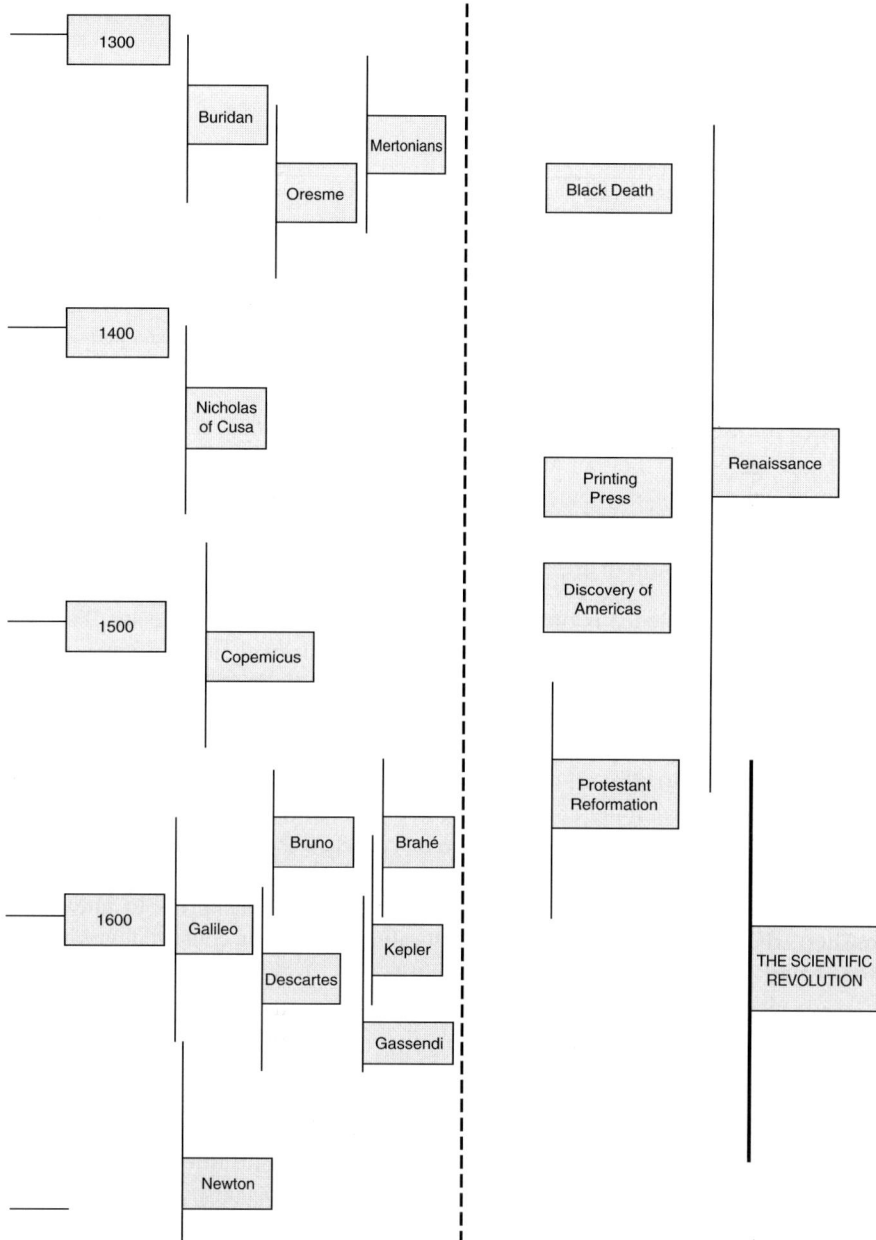

Fig. 5.1 Time-course of some salient contributions to thought and of events related to the secularisation process during the period 1300–1700. The alleged dates of the Scientific Revolution are highlighted

post-Renaissance, post-Reformation, post-printing press society as well as a further evolutionary stage in an ancient intellectual tradition. Table 5.1 summarises some of its main distinctive features. Each feature requires comment.

Table 5.1 Major differences between mediaeval Aristotelianism and the new natural philosophy of the 17th century

Aristotelianism	Natural philosophy
Phenomena explained by reference to God	God not evoked in explanations of the natural world
World viewed as 'organic' and purposive	World viewed as 'non-organic' and purposeless
Mainly qualitative	Fundamentally mathematical
Non-progressive	Inherently progressive
Largely unrelated to technology	Deeply connected with technology

Invoking or Not Invoking God

In our own times, science is often linked with agnosticism and atheism. That was not the case with 17th century natural philosophy, even though it explained the world in naturalistic rather than theistic terms, in stark contrast to mediaeval Aristotelianism. The overwhelming majority of natural philosophers were trained in the old schools of thought and therefore loaded with its metaphysical prejudices, steeped in the need to refer all their discoveries in some way to religion. In the 17th century, and to an extent the 18th, the main thrust of natural philosophy was to support, not to oppose, religion and metaphysics. In the dozens of disputes among the leading thinkers (who were as argumentative as any group of scientists today) the religious fitness of ideas formed the foreground, not the backdrop, to intellectual conflict. All causal explanations were naturalistic, but their consistency with religious belief remained a primary consideration.

Mechanism Instead of Teleology

The sharpest contrast between the thinkers of the Scientific Revolution and those of the Scholastic era lay in their view of the natural world as intrinsically *inanimate, non-organic and purposeless*. As the Aristotelian reliance on 'final cause' in explaining phenomena was abandoned, philosophers ceased to view the world as organic and purposive. Astrology lost its intellectual respectability, and in the hands of Robert Boyle and others, alchemy began to give way to a mechanistic chemistry. This change was fundamental to the establishment of modern science; but it created obvious difficulties for a scientific biology, which we shall discuss in the chapters to follow.

Mathematical Reasoning

Buridan, Oresme and others had advanced mechanistic concepts, but despite their innovations they did not have the mathematical tools (or the cultural ambience) to develop them as did Galileo, Descartes and Newton. For example, the 14th century Mertonians did not, for the most part, relate their mathematical discoveries to descriptions of the observable world. In contrast, one of the earliest declarations of the Royal Society (during the 1660s) was that all philosophical discourse should be as close as possible to mathematics, the most trustworthy language for conveying the unvarnished truth. Aristotle's physics had been largely *qualitative*; but natural philosophy was to be *quantitative* at every possible opportunity.

Inherent Progressiveness

Torricelli's experiment led to an obvious question: *why* does a vacuum form above a sufficiently long column of liquid? Attempts to answer this question led to the concept of air pressure, which suggested further questions. *Why* does air exert pressure? How does the pressure of air (or any gas) relate to the volume in which it is contained, its temperature and its mass? These inquiries ultimately led to thermodynamics, kinetic theory and statistical mechanics. The mechanistic approach to knowledge, exemplified by Torricelli's work, was *inherently progressive*. In contrast, the Aristotelian pronouncement that 'Nature abhors a vacuum' was not progressive at all. A final cause of all matter was to fill or prevent vacua. The statement did not invite, or even allow, further questioning. Aristotelian explanations were compatible with the essentially static culture of mediaeval Europe; but the changing, inherently progressive, increasingly secularised world of early capitalism seemed to demand a changing, inherently progressive mode of understanding such as natural philosophy.

Relationship to Technology

By improving understanding, the natural philosophers believed that they could improve human welfare in practical ways, for health, longevity and prosperity. That was Francis Bacon's idea (Chapter 2), though its roots could be traced to the activities of the Florentine *umanisti* of the 14th and 15th centuries. In common with the *umanisti*, the natural philosophers of the 17th century and later combined practicality with the quest for knowledge. This was perhaps the most radical aspect of the new natural philosophy; i.e. it assumed, for the first time in Christian history, that humans could improve their lot (even create Paradise on Earth) *through their own efforts* rather than through Divine Grace.

Science and Cultural Relativism

We have seen that social context played a fundamental role in the historical emergence of science. Today, the advancement of science depends entirely on the wider society, e.g. funding resources, institutions, cultural symbols and the general *Zeitgeist*. Science is deeply and inextricably embedded in our culture.

Nevertheless, scientific ideas are justified by rigorous and critical sifting without reference to any consideration external to science itself. Therefore, the enterprise of science must be seen as an *objective* exploration of reality. It is a fallacy (technically known as the 'genetic fallacy') to explain current beliefs and ideas solely by reference to their origins. Cultural relativists hold that science is no different from the beliefs of any other culture, but they are wrong. The *practice and direction* of scientific thought and experiments depend on contemporaneous socio-economic and political conditions, but the *knowledge generated* by such thought and experiment is independent of culture. Scientists would stop researching if they did not firmly, and justifiably, believe that they were uncovering real phenomena and mechanisms.

However, these phenomena and mechanisms and relationships are described in the *language* of a particular time and place. Moreover, all scientific accounts implicitly use analogy and metaphor, so they are not literal. (To presume that they are literal is to commit another well-known fallacy, which Whitehead[20] called the 'error of misplaced concreteness'.) Several philosophers of science have been led to adopt this position, known as 'critical realism', in recent years; but practising scientists shift their philosophical viewpoints according to circumstances, without necessarily being aware that they are doing so. We discuss this further in the Appendix.

Taking Stock

Before we move on to explore the origins of *biology* as opposed to science in general, let us reflect on the questions that led us into the historical survey occupying Chapters 3–5. Chapter 1 ended with the inquiry:

> *Why did so curious and 'unnatural' a way of seeking knowledge as modern science arise in 16th and 17th century Europe, but apparently in no other place and at no other time in history?*

We can now answer: the Scientific Revolution occurred under a unique combination of conditions.

On the one hand, European learning in the 16th and 17th centuries inherited an intellectual tradition that stretched back to the naturalistic philosophy of Classical Athens, transmitted and augmented through Alexandria, the Islamic golden age and mediaeval Scholasticism. At each stage in its journey from the time of Plato and Aristotle (and their antecedents in Ionian Greece), this body of learning had been

---

[20] Science and the Modern World, Macmillan, New York (1925).

supplemented with new discoveries in what we would now call mathematics, physics, chemistry, anatomy, physiology and medicine.

On the other hand, secularisation had gained strength by the 16th century. Feudalism had given way to early capitalism, with its emphasis on individualism, technology and progress. The Renaissance had brought learning and practicality together. New lands had been explored and new sources of wealth discovered. The Church no longer exerted monolithic power, and the Reformation had challenged traditional thinking. Every major city had printing presses. Thus, the conditions for naturalistic thought that supposedly had emerged in Classical Greece emerged again in 16th century Western Europe; but this time, the individuals involved were interested in both learning and technology.

Natural philosophy emerged ahead of its competitors during the period in which capitalist culture matured. It proved highly durable and adaptable.

At the end of Chapter 1 we added some supplementary questions:

1. *Why do we consider science more reliable than the more widespread (supernaturalistic, specific, concrete) alternatives?*

Science has been remarkably successful in providing an ever-improving understanding of the natural world. It has progressed concomitantly with improvements in technology that have greatly increased our health, wealth, comfort and longevity. Although the relationship between technology and science is complex (Chapter 2), this coincidence looks like a fulfilment of Bacon's dream. If a 17th century natural philosopher had envisioned our modern world, it would probably have seemed in many ways like Paradise on Earth. It is the success of technology rather than the advancement of knowledge *per se* that makes us consider science so reliable. More fundamentally, if less obviously, science as a way of thinking and knowing has become very deeply embedded in our secularised culture, to the extent that we cannot truly conceive of a credible alternative.

The practice and the perceived social value of science (previously known as natural philosophy) have changed as socio-economic conditions have changed, but – broadly speaking – the status of science as the only reliable source of knowledge has become ever more firmly established. Certainly the knowledge it has generated has never been rivalled in depth and detail, consistency, or explanatory and predictive power.

2. *Could appearances have deceived us? Was there, after all, a precedent for what we now call science?*

We can now answer with certainty: no, there was no such precedent. However, although there was a fundamental change in worldview during the Scientific Revolution, the contributions of the great Islamic and mediaeval European scholars were indispensable for it. The *umanisti* of the Renaissance and, following them, the natural philosophers of the 17th century repudiated those contributions, intent as they were on a 'fresh start'. Many more recent historians of science have followed their example, apparently blindly. 'Science' as we characterised it in Chapter 1 did not exist before the 17th century (in name, not before the 19th), though it had important precursors.

These conclusions also answer the questions we raised in Chapter 2:

Is science uniquely the product of post-mediaeval Western Europe?
If not, where (and when) else did it arise – and why do we never hear about it?
If the naturalistic learning of Classical Greece was not science, how did it come to exert such a profound influence on the emergence of science in Europe 1800 years or more afterwards?
Technology and knowledge developed more or less independently in ancient civilisations, even when they were rooted in the same culture. What made our culture different?

Finally, we have answered the questions raised in Chapter 4:

What contributions did the Muslim world of the 8th–12th centuries make to learning, how did it come to make them, and why are they relegated to little more than footnotes in most histories of science?

The Muslim empire assimilated almost all the regions in which the Classical learning of Athens and Alexandria was preserved; the only major exception was Constantinople. Because early Islam encouraged learning and (subject to monotheistic faith) freedom of thought, a remarkable flowering of intellectual endeavour resulted. This vastly important period of learning has a 'footnote' status in most traditional histories of science because, once again, the pioneering natural philosophers of the Scientific Revolution sought to repudiate the past; and we have been slow to demur.

Chapter 6
The 'Scientific Revolution' in Biology

Accounts of the Scientific Revolution focus on the rise of mechanics, the new mathematical account of the physical world, and the dismissal of Aristotelianism. But we have left open the question of whether there was also a 'scientific revolution' in biology.

The foundations of present-day medicine and biology were laid at the same time as those of mechanics. The pioneering work of modern anatomy, the *Seven Books on the Fabric of the Human Body* by Andreas Vesalius (1514–1564),[1] was published in the same year (1543) as Copernicus's book. During the century that followed, approaches to learning changed fundamentally across the entire range of 'sciences' from mechanics to medicine. But there was a crucial difference. Bacon, Galileo, Descartes and Gassendi rejected the Aristotelian account of the inanimate universe entirely, but Aristotelianism was expunged from medicine and biology – if it has ever been completely expunged – not in a few confrontational decades but after a war of attrition that persisted into the 19th century. Why, and what were the consequences? We shall start to answer these questions, which will dominate much of the rest of this book, in the following pages.

William Harvey (1578–1657), who gave us the theory of blood circulation, was a contemporary of Bacon and Descartes, but he did not share their antipathy to Aristotle; his explicit target was Galen. Harvey's revolution and the background to it illustrate the process by which scientific ideas change. When we compare Harvey's writings about physiology with those of Descartes, we see how the Scientific Revolution was connected to the study of life. At the same time, we start to understand why Aristotelianism persisted in biology and medicine and gave rise to long-lasting debates.

The Demise of Galenism

Galen's Account of Blood Movement

The Greek-born Roman doctor Claudius Galen was a follower of Aristotle, though he lived some 500 years later (*c*. AD 129–216). Physician to three successive

---

[1] One reason why the work of Vesalius became so prestigious was the high artistic quality of the anatomical illustrations, since they had been executed in Titian's workshop. Renaissance art influenced science.

P.S. Agutter, D.N. Wheatley, *Thinking about Life*,
© Springer Science + Business Media B.V. 2008

emperors, he exerted great and durable influence; his works were to dominate medical teaching for some 1,500 years (Chapter 4). He championed an experimental approach to medical investigation. Throughout his life he dissected animals in the quest to understand how living bodies are organised, and he advised doctors to follow his example in order to discover new facts and improve their surgical skills. Most of his anatomical observations are now deemed accurate, though there are a few glaring exceptions. However, his insights into physiology (the way the body functions) were of variable quality. In agreement with modern beliefs, he stated that urine was formed in the kidney, recognized that tuberculosis was contagious and that rabies may be spread by dogs, and established that arteries carry blood, not air. He also identified the heart valves. But his account of the movement of blood around the human body differed radically from our modern view (i.e. Harvey's view). He held that:

1. The blood in the arteries is separate and distinct from that in the veins. The liver generates venous blood from nutrients digested by the intestine. Venous blood, with its cargo of nutrients, is then consumed by the body's tissues. Arterial blood, in contrast, transports vital spirits throughout the body.
2. The pulmonary vein (now known to return blood from the lungs to the heart) is *not* a blood vessel. It supplies pneuma/spirits to the arterial blood while the pulmonary artery evacuates 'sooty vapours'.
3. The heart works by sucking blood out of the veins, not (as we would now say) by pumping it into the arteries. Galen said that arteries 'agitate' or mix the blood with the vital spirits rather than moving the blood from place to place.
4. Blood from the veins could seep from the right (venous) to the left (arterial) side of the heart through invisible pores in the septum that divides the ventricles[2] (Fig. 6.1).

Anomalies in Galenism

Although Galen's account was taught in medical schools throughout Europe during Harvey's lifetime, there was growing discomfort about it. Some conflicting evidence dated back to the Islamic golden age (Table 4.1). Rhazes and Averroës had both criticised Galen – Rhazes may have known about valves in veins – and Ibn al-Nafs had described the pulmonary blood transit in something akin to modern terms. (Although the main texts used for medical teaching were those of Rhazes and

---

[2] Several of these proposals were originally due to Aristotle. For example, Aristotle had said that food was converted to blood and distributed around the body, and that the function of the lungs was to cool the body; he had also grasped the function of the kidneys more or less correctly, dismissing the old belief that urine is formed in the bladder. This leads to an interesting question. Harvey's great work was written at a time when the intellectual reaction against Aristotle had achieved maximum momentum. Yet the discovery of the circulation was taken by Harvey himself, and by others, as a refutation of Galen, *not* of Aristotle. Why? Part of the reason was that Galen, not Aristotle, was accepted as the founder of traditional medical teaching and practice, but that is not the whole explanation (discussed further in Chapter 7).

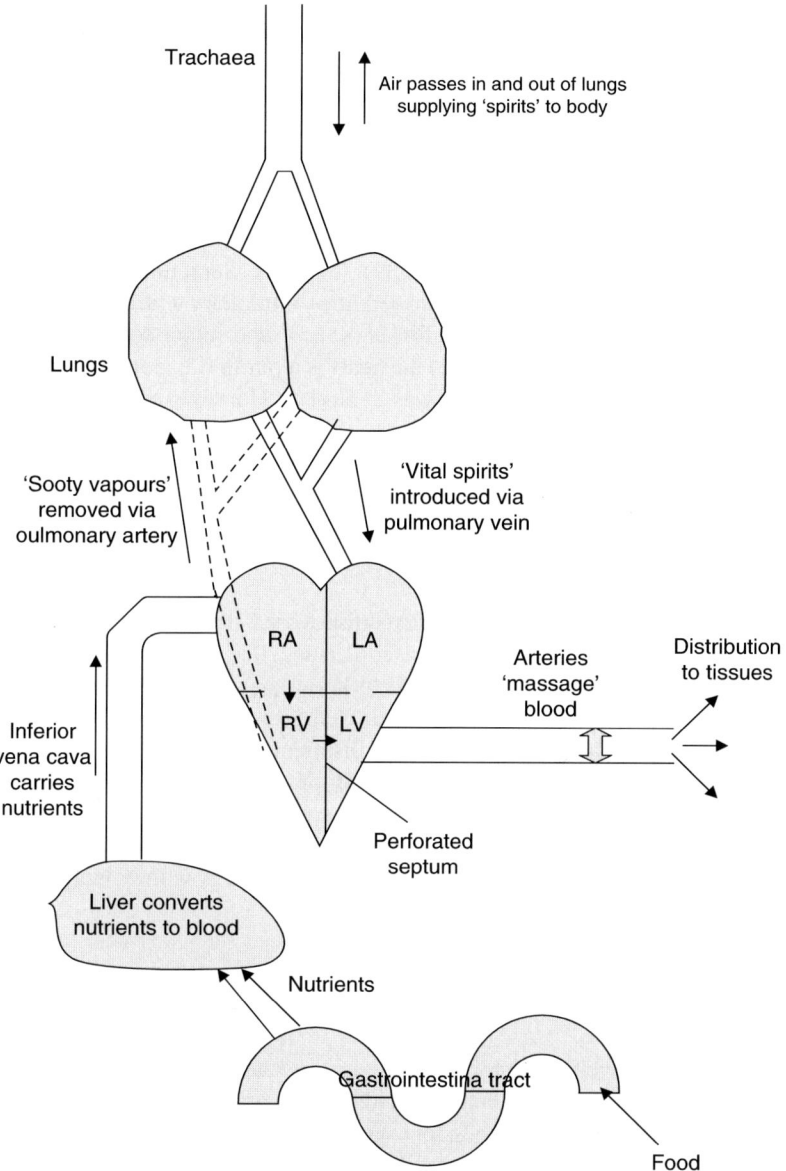

Fig. 6.1 Simplified diagram of mammalian (including human) blood movement as envisaged by Galen. Nutrients pass from the gastrointestinal tract into the liver where they are converted to blood. The inferior vena cava arises from the liver and conveys blood to the right atrium (RA) of the heart. 'Sooty vapours' are removed via the pulmonary artery, and the nutritive blood passes into the right ventricle (RV) and then through the pores in the septum to the left ventricle (LV). Here it is supplied with vital spirits conveyed from the lungs to the left auricle (LA). The arteries then carry the blood containing vital spirits to the tissues, 'massaging' it by their pulsatile action and thus drawing the blood out of the heart

Avicenna, their contents were often attributed to Galen, perhaps an illustration of Renaissance bias.) During the late Renaissance, Leonardo da Vinci had attempted to rethink the process of blood movement, showing that Galen could be questioned. Most of the anomalies, however, accumulated after the dawn of the Scientific Revolution.

Vesalius found that the inferior vena cava, the great vein that Galen believed to carry nutrients from the liver for distribution to the rest of the body, does *not* originate in the liver. He also showed that there are no pores in the septum between the ventricles. Realdo Colombo (1510–1539), Michael Servetus (1511–1553) and Andrea Cesalpino (1519–1603) described the pulmonary circulation – perhaps independently, perhaps plagiarising Ibn al-Nafs – and understood how the heart valves operate. Colombo realised that the heart is a pump (i.e. not a suction device). But the discovery of *venous valves* was central to Harvey's argument and to the change in thinking that followed it.

The Discovery of Venous Valves

No one is quite sure who in Europe first 'discovered' valves in veins, or whether these 'discoverers' deliberately plagiarised Rhazes. Perhaps the question of precedence is unimportant. It seems sufficient to know that the European 'discovery of valves in veins' began in the 1530s but was dragged out over almost a century.

Not surprisingly, Harvey himself named his mentor Fabricius as their discoverer, but he acknowledged the contrary opinion of his lifelong Parisian opponent Riolanus.[3] There is no doubt that Fabricius's exquisite description of 'membranous portals' was a crucial influence on Harvey, though Fabricius believed that the valves did not close completely, so he never understood how they work.[4]

Parallelism with the Demise of Aristotelianism in Mechanics

In Chapter 5 we described the anomalies that accumulated in Aristotle's physics before the early 17th century. Those anomalies included some from the Islamic golden age (the studies of al-Biruni and Averroës, for example), some from mediaeval Scholasticism

---

[3] In *Exercitatio anatomico de mortu cordis et sanguinis in animalibus* (1628), Harvey wrote: '*Fabricius, [or, as Riolanus would have it, Jacobus Sylvius] did first of any delineate those semilunary membranous portals…*'. Riolanus (Jean Riolan) was the doyen of Parisian surgery and medicine, and Harvey's contemporary. A scion of eminent Galenist teachers on both sides of his family, he was implacably opposed to Harvey.

[4] We should not, with hindsight, blame Fabricius for his error. Unless our minds have been primed with an idea of *bulk* venous flow, we could have no reason to suppose that the valves would open and close. Fabricius, like all his peers, saw the body through Galenist eyes, so his mind was not primed in that way.

(Buridan, Oresme, Nicholas of Cusa) and many more from the period of the Scientific Revolution (Copernicus, Brahé, Kepler, Galileo, Descartes, Gassendi). Only after that prolonged process did an entirely new way of thinking about physics evolve (i.e. Newton's). In the present chapter we have shown a similar accumulation of anomalies in Galen's account of blood movement, i.e. during the Islamic golden age (Rhazes, Avicenna, Ibn al-Nafs), the later mediaeval period (Leonardo da Vinci) and (especially) the Scientific Revolution (Vesalius, Colombo, Servitus, Cesalpino and the discoverer(s) of venous valves). Only after that protracted build-up did an entirely new way of thinking about human physiology evolve (Harvey's).

The parallel is striking. Does it point to a general 'mechanism of theory-change' in the history of science?

Is There a General Process of Theory-Change in Science?

During the third quarter of the 20th century, Thomas Kuhn observed that events in the history of physics follow a previously unsuspected pattern. Most of the time, physicists agree about the assumptions under which they are working and the methods they use. They explore phenomena in the context of a shared body of belief, adding details to knowledge, occasionally elaborating a theory or finding new applications of it. That is 'normal science'. Sometimes, however, anomalies accumulate – results that do not fit comfortably into the accepted scheme of things – and a 'crisis' ensues, causing a period of 'extraordinary science', a revolutionary change in the way scientists construe the world. Such a revolution results in new assumptions and methods, a new shared body of belief.[5] A new period of normal science then follows, but only after the most prominent adherents of the old way of thinking have ceased to work and have been succeeded by a younger generation who accept the new viewpoint.

The revolution that replaced Aristotle's physics with Newton's mechanics, and the twin revolutions that 'replaced' Newtonian mechanics with relativistic and quantum mechanics, illustrate Kuhn's account of the history of science. Harvey's revolution against Galenic physiology is another example. To accept Harvey's theory was to deem Galen factually and philosophically wrong. But the Galenist 'old guard', notably Riolanus, could not be persuaded. When Harvey's great work was published, he apparently lost many patients. He certainly provoked vitriolic dissent, which subsided only after the death of Riolanus and other leading Galenists of his generation.

Kuhn's study proved that science does not progress by straightforward accumulation of data and successions of illuminating discoveries. The history of science is not a

---

[5] Kuhn used the often misunderstood word 'paradigm' for this conjunction of assumptions, beliefs and methods. He chose the term because he was able to trace the assumptions, beliefs and methods of any normal science period to some pioneering landmark experiment or set of observations. Unfortunately, the use of the word in Kuhn's seminal essay and its successors was somewhat flexible, so we prefer to avoid it.

one-route journey from the darkness of ignorance to the light of knowledge and truth. It is a messy, haphazard process, inseparable from the rest of culture, and the directions it takes can only be traced by retrospective reconstruction.

Harvey's Theory of the Circulation

Harvey was exposed to the newly-evolving metaphysic of natural philosophy from his formative years and was a strong proponent of the Baconian approach to experiment:

> I profess to learn and teach anatomy not from books but from dissections... not from the tenets of Philosophers, but from the fabric of Nature.

His great work, *Exercitatio anatomico de mortu cordis et sanguinis in animalibus (DMCS)*, used the approach of natural philosophy to study a *biological* process, the circulation of the blood.

In the first part of *DMCS*, Harvey disproved the four points of Galenist 'physiology' described above. Like Galileo (Chapter 5), he combined evidence from observations and experiments with clear reasoning from principles. He showed that all the venous valves opened in the direction of the heart, not necessarily against gravity. A solid probe could be passed through a vein in one direction, but not in the other because the valves stopped it. By ligating the forearm, he showed that segments of vein between valves could be emptied of blood then refilled when the ligature was removed, proving that the valves could close completely. He performed experiments on live frogs. He showed that the heart is muscular. Then he calculated that the heart pumps more blood in an hour than is contained in the entire body, proving that the blood must be 'recycled'.

Together, these discoveries showed that the blood motion imagined by Galen was impossible, so the orthodox teaching of the previous 1,500 years was invalid. Harvey was led to the radical inference that the blood does not merely flow but flows in only one direction (Fig. 6.2):

> Since all things, both argument and ocular demonstration, show that the blood passes through the lungs and heart by the force of the ventricles, and is sent for distribution to all parts of the body, where it makes its way into the veins and porosities of the flesh, and then flows by the veins from the circumference on every side to the centre, from the lesser to the greater veins, and is by them finally discharged into the vena cava and right auricle of the heart, and this in such a quantity ... as cannot possibly be supplied by the ingesta ... it is absolutely necessary to conclude that the blood in the animal body is impelled in a circle, and is in a state of ceaseless motion.

Connections with the Revolution in Mechanics

Kuhn said that only when radical scientific claims are *timely* can they undermine the consensus viewpoint. Harvey's claim was 'timely' in at least two respects. First, as we have seen, anomalies had accumulated in Galenism during the previous

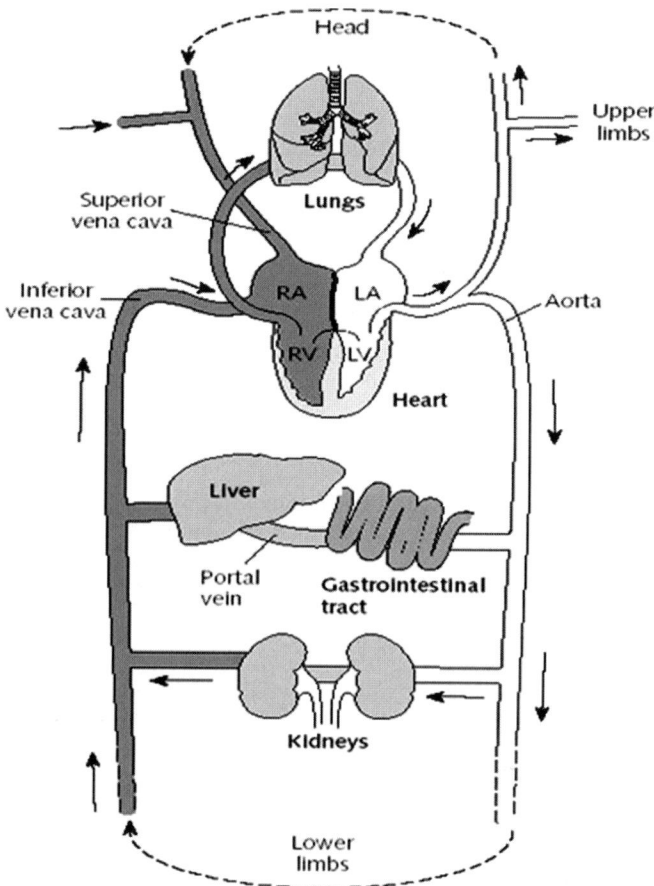

Fig. 6.2 Simplified diagram of mammalian (including human) blood circulation as we understand it today. Blood enters the right atrium (RA) of the heart from the two great veins, the superior and inferior venae cavae. It then enters the right ventricle (RV) to be pumped at low pressure through the lungs. Oxygenated blood from the lungs returns to the left atrium (LA) and is pumped by the left ventricle (LV) at high pressure through the aorta, then through the arteries of the systemic circulation, and via the capillaries into the veins before returning again to the right heart. The unshaded vessels carry oxygenated blood (to the tissues) and the shaded ones carry deoxygenated blood (from the tissues). This was essentially Harvey's conception, amply corroborated during the past 300 years

century, and the accumulation of anomalies presages a scientific revolution. Secondly, the general idea of 'cycle'/'circulation' was being assimilated into the new worldview of natural philosophy when *DMCS* was conceived.

This second point deserves comment. The revolutions in mechanics and physiology were parallel, but it seems that they were also *connected* by intellectual cross-currents.[6]

---

[6] The conceptual juggling by which Galenists sought to comprehend anomalies such as venous valves can be compared to the increasingly complex pattern of deferents and epicycles by which Ptolemaic astronomers sought to accommodate new data before the Copernican model was finally accepted. The comparison, though inexact, is suggestive.

When Harvey went to study medicine in Padua in 1597, the new astronomical ideas initiated by Copernicus had become controversial throughout academic Italy. Giodarno Bruno was burned at the stake in Rome in 1600 for - among other heresies – speaking about the 'cycle' of the planets, and the incident must have excited debate in all the Italian universities. Moreover, Bruno had surmised that: *'In us the blood continually and rapidly moves in a circle'.*[7] While Harvey was a teenager at Caius College, Cambridge, he may have encountered this controversial opinion and been inspired by it; Bruno had spent some years in England in the 1590s. Significantly, in the dedication of *DMCS* to Charles I, the king is likened to both the sun at the centre of the universe and the heart at the centre of the body:

> To the heart is the beginning of life, the Sun of the Microcosm, as proportionally the Sun deserves to be called the heart of the world, by whose virtue, and pulsation, the blood is moved, perfected, made vegetable, and is defended from corruption, and mattering; and this familiar household-god doth his duty to the whole body, by nourishing, cherishing, and vegetating, being the foundation of life …

The 16th–17th century Scientific Revolutions in physics and biology/medicine were not only concurrent, and not only parallel, but actually interlinked. This would seem to justify the claim that physics and biology started to become 'scientific' at the same time in history and as a result of the same cultural processes. Yet Aristotelianism survived in biology and medicine long after it was dead in physics. We need to investigate this seeming contradiction.

Conceptions of Natural Philosophy

The Scientific Revolution led to more than one conception of how natural philosophy should be pursued. The three 'major prophets' of the new way of thinking were Galileo, Descartes and Francis Bacon (1562–1625).

Bacon had little grasp of mathematics, and unlike Galileo and Descartes he made virtually no lasting scientific contributions of his own. His legacy was a prescribed *method* for obtaining knowledge. In Bacon's view, we must rely on tradition and authority in matters of law, religion and politics,[8] but true understanding of the natural world must be based exclusively on meticulous observation. Natural philosophers must collect observational data without personal prejudice or regard for authority or popular belief. Apparent patterns in the observations must be inspected carefully for inconsistencies or counter-instances. No prior belief or expectation must be allowed to

---

[7] This quotation, and much of the evidence on which we base this section, comes from Pagel W (1951) Giordano Bruno, the philosophy of circles and the circular movement of the blood. *J Hist Med* 6:116–124, and Pagel W (1957) Philosophy of circles – Cesalpino – Harvey. A penultimate assessment. *J Hist Med* 12:140–157.

[8] He was a brilliant lawyer who became a Member of Parliament at a young age and subsequently held the post of Lord Chancellor, until his fraudulent practices came to light and he was 'allowed to retire'.

interfere; natural philosophers must behave like ideal lawyers preparing cases. Only when there was copious evidence could tentative conclusions be drawn with generalisations needing to be based on large numbers of particular instances. Reliable reasoning in natural philosophy was therefore *inductive* (from the particular to the general). Bacon was as much an *empiricist* as his 13th century namesake and the other great Scholastic thinkers; though, unlike them, he sharply dissociated knowledge of the natural world from theology. Bacon's prose was erudite, concentrated and tough, his style polished and uncompromising and his rhetoric powerful.[9] The 'Baconian method' became institutionalised in British natural philosophy a generation or so after its author's death with the founding fathers of the Royal Society pronouncing it the only acceptable route to knowledge. Its influence spread internationally.

Galileo's work on mechanics places him among the greatest of western thinkers, and it showed the need for both observation and reasoning in natural philosophy. In Chapter 5 we described his demonstration that the moon reflects light, not *despite* but *because of* its rough surface. This 'proof' required *induction* from careful and unbiased observations, coupled with *deduction* from a principle governing the reflectiveness of surfaces. The example shows that, in practice, natural philosophy (science) requires both the evidence of the senses and reasoning from general principles.

Descartes, whose main aim was to become 'the new Aristotle', emphasized the 'reasoning' aspect of natural philosophy. He sought to reconstruct the whole of human knowledge from scratch, building it like Euclid's geometry, through a succession of theorems *deduced* from a few apparently unchallengeable axioms. In other words, he held that the wellspring of reliable knowledge is human reason, a philosophical position known as *rationalism*. Descartes was just as 'scientific' as Galileo and Newton: they all agreed that experiment and observation were the indisputable basis for knowledge about the world, but Descartes emphasised the central importance of *deduction* from incontrovertible general truths as opposed to *induction* from particular instances.

Descartes was a builder of conceptual systems, Bacon a prophet of method, Galileo an inspiring teacher who led by example. Of the three, only Descartes made any substantial pronouncements about biology and medicine.[10] His correspondence with Harvey foreshadowed a philosophical debate that was to echo through biology and medicine for the next two centuries.

---

[9] For example, he called the various possible errors of reasoning (e.g. jumping to conclusions, ignoring counter-instances, believing in ancient authorities such as Aristotle) 'idols' of the mind, the market-place and so on. The word 'idol' had great rhetorical force when Bacon wrote his first version of the *Novum Organum*, around 1610. In Protestant England, 'idolatry' was a pejorative term for Roman Catholicism, which was not only despised and hated but also feared: the Gunpowder Plot was a very recent memory. To Bacon's Anglican contemporaries, 'idol' therefore implied treason, terror, darkness, misery and a return to the benighted past. Bacon knew how to influence a jury.

[10] Descartes wrote his *Treatise on Man* around 1630, well before his famous *Discourse on Method*, but it was not published during his lifetime. Apparently he feared that his mechanical explanations of living functions would expose him to a charge of heresy. Indeed, he seems to have been so afraid of condemnation that his book is ostensibly about a 'perfect robot replica of a human being with no soul' rather than a human being *per se*.

Harvey and Descartes

Descartes famously declared that animals are 'mere mechanisms'. Fundamentally, they are no different from inanimate objects, so everything about them should be explained in terms of the science of mechanics. Only in humans is there a 'rational soul', a subject for theology rather than physics. This recalls the Augustinian ('Platonic') division between the secular and the eternal worlds; mechanism operates in the former while the autonomous, rational soul belongs to the latter.

Descartes was fascinated by the mechanical operation of the body, so he was greatly attracted by Harvey's circulatory theory as a system of tubes, valves and pumps. He seems to have understood it immediately, which many Galenists did not. In return, Harvey wrote positively and approvingly of Descartes. Nevertheless, these two highly influential thinkers differed in their views of physiology.

Descartes accepted Harvey's concept of 'blood circulation' but contrived to retain many Galenist notions as well, and also to introduce errors of his own. Primarily, he refused to recognise the heart as an active pump; instead, he agreed with Galen (and Aristotle) that the heart operates primarily as a source of *heat*. When cold blood enters the heart it is heated by friction. Thereupon it vaporises, expanding the heart walls, opening the valves into the arteries and closing the valves by which it entered. The escape of this heated blood into the arteries creates the pulse. When cold blood re-enters the heart chambers it makes the walls contract, enhancing the frictional heating. This account of heart action was more overtly 'mechanical' than Harvey's, but within little more than a generation it had been rejected. 'Mechanical' is not synonymous with 'true', and great thinkers are not always right. But for a time at least, opposition to Harvey's theory did not come exclusively from Galenists; it also came from Descartes.

On one level, the difference between Harvey and Descartes was technical and factual so it was fairly easy to resolve. But there was also a deeper difference, which heralded a persistent (or recurrent) schism in biology and medicine. Harvey seems to have believed, with Aristotle, that organisms are endowed with inherent purposiveness, and that the self-moving, circulating blood embodies that purposiveness in animals. (Indeed, he reiterated Aristotle's dictum that *'Nature does nothing in vain'*.[11]) In other words, he saw a fundamental distinction between living and non-living mechanisms. Descartes considered that exactly the same mechanical principles would suffice to explain both the animate and the inanimate.

The difference was not quite as simple or clear-cut as that. For example: both Harvey and Descartes knew that nerves somehow carry messages from the brain to the body and vice versa. Descartes gave a good physical description of nerve fibres and thought about the means by which they can be stimulated. On the other hand, he believed that the rational soul, seated in the pineal gland, controls nerve function.

---

[11] In *DCMS*, he wrote: '… from the symmetry and magnitude of the ventricles of the heart and of the vessels entering and leaving, (since Nature, who does nothing in vain, would not have needlessly given these vessels such relatively large size)…'

That was how he 'explained' the control of our bodies' mechanical actions by free will. His speculation about what we now call 'nerve impulses' betrays an even more peculiar blend of natural philosophy with antiquated belief. He suggested that the fastest particles carried by the blood enter the brain and are directed by the rational soul into the nerves, which he believed to be hollow. They are separated into '*a very subtle flame or wind*', which passes through the nerves into the muscles and inflates them. A synonym for this '*flame or wind*' was '*animal spirits*'. 'Animal spirits' was an explicitly Aristotelian idea (see Chapter 7). Thus, Descartes combined a mechanical interpretation of muscle action – inflation by wind passing through the hollow nerves – with Aristotelianism.

In that respect at least, Descartes was as much in thrall to the influence of Aristotle as was Harvey or any other contemporaneous writer about biology and medicine.

Implications of Harvey's Theory: Testing the Predictions

The gifted vivisectionist Richard Lower (1631–1691), inspired by Harvey's work, performed the first blood transfusion in western history in 1666. Three years later he gave a Harveyan account of the anatomy and physiological action of the heart; a specifically *living* mechanism. Within a decade of Harvey's death, therefore, both medical practice and teaching were beginning to change in response to the circulatory theory.

The theory posed new questions. If venous blood is not formed in the liver, as Galen had said, how *is* it formed? Since blood is passed through the lungs during each transit around the body, what does it do there? If blood does indeed carry nutrients to the tissues, but is not actually made from those nutrients, how do the nutrients enter it? These new questions led to whole new areas of research, just as Torricelli's vacuum experiment (Chapter 5) led to new areas of research, and in time they produced (e.g.) our modern understanding of respiration and digestion. Harvey's science, like Galileo's and Torricelli's, was *progressive* whereas Galen's was not, despite Galen's own insistence on observation and experiment.

However, Harvey's theory made one particular critical prediction,[12] which was not tested until after its author's death. If the blood circulates (heart to arteries to veins to heart), then it must somehow pass from the finest divisions of the arteries into the finest divisions of the veins through pores or connections – '*porosities of the flesh*' – that Harvey himself was unable to see. In other words, the theory predicts the existence of microscopic blood conduits in all tissues. In 1661, capillaries[13]

---

[12] The model proposed by Descartes entailed the same prediction, of course.

[13] The circulatory model could in principle have been reconciled with a system that allowed blood to 'leak' through the tissues from the smallest arteries to the smallest veins, though this may have been difficult given the overall speed of blood movement through the system. The discovery of capillaries proved that the 'porosities of the flesh' were in fact vessels, not open spaces, so the blood is contained within a continuous system of tubes throughout its cyclic journey.

were demonstrated, initially in lungs, by Marcello Malpighi (1628–1694), using a new instrument: the compound microscope.

The Early Microscopists

Like the telescope, the compound microscope was a product of the master lens grinders of the Netherlands (Fig. 2.2); and like the telescope, its revelations transformed our understanding of nature. Microscopy was pioneered during the second half of the 17th century, the time of Newton and the early years of the Royal Society.

Robert Hooke (1635–1703), the Royal Society's curator of experiments, was wholeheartedly Baconian. Besides his many contributions to physics (which brought him into conflict with Newton), he was a noted inventor. Among his technological innovations were the universal joint now used in motor vehicles, the iris diaphragm, a respirator, the anchor escapement and the balance spring that made clocks more accurate, and new or improved barometers, anemometers and hygrometers. In true Baconian fashion, he saw each of these accomplishments as an *application* of natural philosophy. The same attitude inspired his improvements to the microscope. Many years of study and discovery were revealed in the illustrations of his *Micrographia* (1665) (Fig. 6.3). These included the coloured rings that formed when sheets of mica were pressed together, an observation that interested Newton (such coloured spectra are now known as 'Newton's Rings'). Hooke also noted that hairs from the beard of a goat would straighten when dry and bend when wet, a discovery

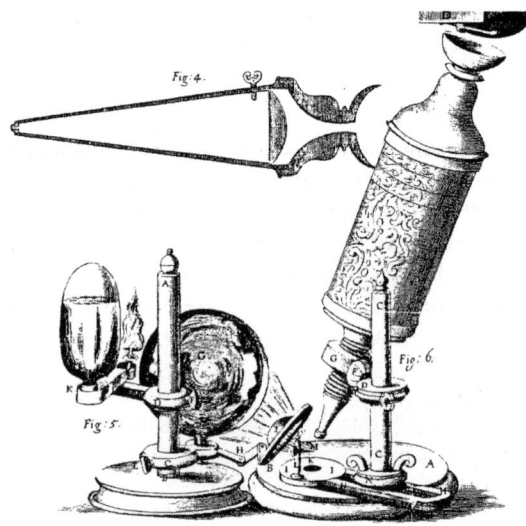

Fig. 6.3 Hooke's microscope, from his *Micrographia*

that he used in the construction of his hygrometer. He examined silk, and wondered whether artificial silk could be spun from 'glutinous substances'. More famously, from his study of the structure of cork, he coined the word 'cell'.

The Dutch textile merchant Antonie van Leeuwenhoek (1632–1723) probably encountered magnifying glasses during his apprenticeship; they were used for quality control in the textile trade, to determine thread densities. When he visited London in 1668, Leeuwenhoek apparently saw a copy of Hooke's *Micrographia*, which included pictures of other textiles besides silk. Thereafter he constructed his own microscopes, which were single-lens devices capable of remarkable magnification. From 1673 he began to send reports of his observations to the Royal Society, of which he was elected member in 1680, and the association continued for the rest of his life. These reports included microscopic descriptions of insect parts, fungi, and tiny organisms in rainwater. Some of these organisms were no doubt protozoa, but they also seem to have included bacteria. Leeuwenhoek's letter announcing this observation met with scepticism, but Hooke confirmed his observations. His other discoveries included blood cells[14] and spermatozoa in many different types of animals, and he concluded, in agreement with Harvey (see Chapter 7), that fertilisation occurs when a spermatozoon penetrates the egg.

The discoveries and pioneering approach to research of the Italian doctor Marcello Malpighi aroused suspicion and hostility among his jealous colleagues. His microscopic studies were informed by the natural-philosophical dictum *'Be prepared to reason, but never go beyond the facts'*. In one of the earliest of these studies, Malpighi announced his famous discovery of blood flow in capillaries in mammalian lung tissue, which confirmed the key prediction of Harvey's theory (see above). He went on to publish detailed microscopic studies of many different human and other animal (and plant) tissues, laying the foundation for the science of histology.

Jan Swammerdam (1637–1680) developed new techniques for preserving and dissecting specimens for microscopy, including wax injection to make it easier to view blood vessels. He may have observed red blood cells before Leeuwenhoek described them to the Royal Society. Swammerdam made pioneering studies of muscle contraction, showing that it depended on a nerve supply but did not involve 'inflation' by air or water intake, thus refuting the account by Descartes (see above). He also gave the first comprehensive descriptions of insect life-cycles.

This period saw other pioneering microscopists, all contributing new information that transformed the way in which living nature was perceived. Such studies remind us that *detailed observation* as well as experiment is crucial for the development of biology. They also illustrate an important general point about the history of science: *the progress of knowledge depends at least as much on new methods and techniques as on new ways of thinking*.

For about a century after this initial burst of enthusiasm, the microscope became more of a toy than a scientific instrument; technical problems with the optics prevented

---

[14] In a letter of 1674, he wrote: '*The red globules of the blood I reckon to be 25,000 times smaller than a grain of sand*'.

further reliable observations. In 1827, J.J. Lister overcame the most serious of these problems by developing a lens that eliminated chromatic aberration (a 'rainbow' outline around the image of a specimen). A new burst of microscopic discoveries followed, leading – among other things – to the first articulation of cell theory (see Chapter 9). In the late 1870s, Carl Zeiss eliminated spherical aberration and introduced oil-immersion lenses, and there was a further burst of discovery, including the first descriptions of cellular organelles (Chapters 10 and 13), demonstrating that new techniques can indeed lead to new knowledge.

The Historical and Philosophical Context of the Early Microscopists

The 17th century pioneers of microscopy were partly motivated by (a) the contemporaneous fascination with *light* (manifest in Vermeer's paintings,[15] Wren's architecture, the optical theories of Huygens and Newton, and the natural-philosophical insistence on the evidence of the senses – mainly vision – as the basis of all reliable knowledge of Nature); and (b) the desire to discover the atomic, particulate basis of all matter, which thanks to Gassendi was a postulate of mechanistic philosophy. But the early microscopists dramatically failed to find 'atoms' of living matter. Instead, they discovered that living things continued to show exquisite and complex beauty no matter how much they were magnified,[16] testifying to the awesome magnificence of God's creation. As we remarked earlier, religious considerations belonged to the foreground, not the backdrop, of 17th century natural philosophy.

Swammerdam's studies are a case in point; they were pious activities. For him, as for the leading Islamic philosophers of earlier centuries, studying the Earth's creatures revealed the greatness of God. That commitment did not affect the objectivity or reliability of his descriptions (or his methods), but they coloured his interpretations and his overall view of nature. For example, he rejected spontaneous generation because he was sure that God would not permit such a random, haphazard process in the world He regulated. By 1675, Swammerdam had come to believe that his studies were undertaken only to satisfy his own curiosity so they were no longer in the service of God. He therefore largely renounced natural philosophy.

Such attitudes would become much less common a century later, when la Mettrie, Hume and others could consider atheism to be philosophically and personally acceptable and major debates in the philosophy of biology were opened. To understand those debates, we need to focus on the question we raised at the start of this chapter: why did Aristotle's biology remain influential long after his physics had been rejected?

---

[15] Vermeer was a close friend of Leeuwenhoek. Vermeer's most famous paintings are believed to have been executed with the aid of the camera obscura that Leeuwenhoek invented and gave to him.

[16] Examine the edge of a new razor blade under the microscope and you will see a rugged outline like a mountain range. Examine the edge of a fly's wing and you will see a perfectly smooth outline. A 17th century observer comparing the man-made 'clean edge' with the God-made one could only have drawn one inference.

Chapter 7
Aristotle's Biology

Even a superficial reading of Aristotle's *Physics* shows how fundamentally our beliefs about the inanimate world differ from his. The Scientific Revolution overthrew his conception of the cosmos entirely. In stark contrast, his surviving works on biology give an overall impression of modernity.[1] Biologists today think about organisms much as Aristotle did, asking similar questions and, in many cases, giving broadly similar (though much more detailed) answers. There *are* important differences, which came to be the battlegrounds in the 'war of attrition' against Aristotelianism that extended from the late 17th to the early 20th century. But the similarities are more immediately striking.

Starting from Scratch

Aristotle inherited a tradition of naturalistic thought governed by logic and reason and studied under the most eminent of mentors; but as far as we know, he did not inherit any tradition of 'biology' beyond the practical, unsystematic knowledge derived from agriculture, fishing, hunting, medicine and everyday human life. In other words, he knew that there is an enormous diversity of plants and animals, which come into being, often (but apparently not always) as a result of sex, then mature, reproduce and finally die; and while they are alive they take in nutrients and excrete waste. Animals breathe; some of them generate heat; most animals move from place to place but some do not, and plants do not. They all have complicated structures, both externally and internally, but the internal structures bear no immediately obvious relationship to activities such as reproducing, using food, producing excreta; or, above all, to being alive rather than dead.

Imagine that your knowledge was equally limited. How would you construct an overall understanding of biology – a systematic, rational set of descriptions *and* *explanations* – with no textbooks, no sources of expertise or authority and no

---

[1] Aristotle made a number of statements that we now know to be false, such as the claim that a lion has a single bone in its neck and that male mammals have more teeth than females, but in the context of his work as a whole those errors are trivial.

P.S. Agutter, D.N. Wheatley, *Thinking about Life*,
© Springer Science+Business Media B.V. 2008

technology such as microscopes to help you? How would you frame questions so that the answers could be obtained from observation and then woven into a comprehensive, naturalistic, coherent, parsimonious understanding of living nature? Very few of us would know how or where to start. But Aristotle did.

Historia Animalium

This book was the first attempt at an organised, comparative, naturalistic description of animals and their parts. Aristotle *classified* several hundred different types of animals on an objective, naturalistic basis, using evidence from comparative anatomy, behaviour, reproduction, habitat and lifestyle. In other words, he used a wealth of observations relating to what are now called anatomy, ethology, reproductive biology and ecology, and from these he constructed a *taxonomy* – much simpler than our present-day classification of animals, but rational. *Historia Animalium* is a marvel of inductive reasoning. Aristotle unerringly identified the most informative evidence, and he knew just when and how to generalise. By applying the principles of reasoning that he had learned from Plato to the world of living animals, he discovered that organisms form natural, hierarchically-related classes. He imposed conceptual order on a world of bewildering diversity.

The book begins with a few 'obvious' remarks. For example, some individuals resemble each other in all their parts (legs, eyes, teeth etc.): they belong to the same *species*. Other individuals resemble each other more distantly; they are broadly similar, but their parts differ in detail (long or short legs, eyes at the front or the sides of their heads, long canine teeth or large molars). They may belong to the same *genus*.[2] The most general groups are animals with blood and animals without blood, categories that are more or less identical with vertebrates and invertebrates, respectively. Aristotle recognised that apes are very similar to humans, different 'species' but the same 'genus', and that some animals-with-blood that live in the sea are not fish but mammals.

Another 'obvious' remark at the beginning of the *Historia* is the division of anatomical parts into 'simple' and 'composite'. A leg is composite: if you cut it in half, the halves are not identical. But if you remove a bone from the leg (the femur, for instance) and cut it in half, the two halves *are* identical, more or less; they are both bone. This observation may seem trivial, but Aristotle used it to infer a profound generalisation: *all composite parts are built from different combinations of the same simple parts*. Out of a limited set of building blocks you can construct a wealth of different structures. In modern science we speak of a limited set of chemical elements combining to form an unlimited wealth of compounds, a limited set of cell types

---

[2] Aristotle used this word differently from us; we would say *class* (e.g. mammals, birds, fishes) or *order* (e.g. carnivores, insectivores, rodents).

combining to form an unlimited wealth of organism types, and so on.[3] From a simple observation, Aristotle had wrung what we now regard as a deep scientific truth.

In one short book, a descriptive framework for animal biology was constructed from scratch. But that was only the start. In his other biological works, Aristotle sought to discover the *causes* of organisms, their parts and their habits.

De Partibus Animalium

Historia Animalium describes; *De Partibus* explains. But Aristotle's attempt at physiology, the ways in which anatomical structures perform the activities of life, was hampered by the complete absence of experimental techniques. He could only dissect dead animals and make external observations of living ones, so it is hardly surprising that many of his inferences were mistaken. Typically, he began by reasoning from simple observations. For instance, animals have to eat and excrete to stay alive, so the most important parts must be those involved in nutrition and excretion. Unfortunately, he had no means of analysing the processes of digestion and absorption, and he could not begin to understand what the liver does. He believed that food was transformed to heat, which was then carried via the blood to nourish the body, the basis of Galen's beliefs about blood movement (Chapter 6). It was a considerable achievement to have grasped the workings of the kidney, another aspect of Aristotle's physiology echoed by Galen 500 years later. On the other hand, his account of respiration was quite different from ours: he considered it to be the means by which land animals acquired *pneuma*, by virtue of which they were living rather than non-living (see the section on *De Anima*, below).

De Partibus contains one very important general principle: all biological structures have functions, and the function is the *point* of the structure. Although Aristotle's ideas about functions were perforce speculative (and in many cases wrong), he never seems to have doubted that principle. In his own words: '*Nature never makes anything that is superfluous*'. The principle is important because it leads to an understanding of 'cause' in biology.

In effect, the four causes (see Chapter 3) are arranged into two pairs. The material and efficient causes together explain *how* a structure (or behaviour) is produced; the formal and final causes together explain *why* it is produced. The 'why' takes precedence over the 'how'. The form of a lung, or a kidney, or an eye, or a leg, is determined by the *use* or *purpose* of that organ. The means by which the animal produces that organ, and the material from which it produces it, are secondary considerations. Thus, Aristotle's approach to biology is fundamentally *teleological*: to understand the *cause* of an object, the first requirement is to understand its *purpose*.

---

[3] By similar reasoning, Aristotle believed that everything in the world is made of different combinations of the four elements of Empedocles (earth, water, air and fire). For many years after the Scientific Revolution, the pioneers of chemistry wrestled with the problems implicit in this claim, and not until the early 1800s was our modern sense of 'element' established (see Chapter 8).

The part functions in the context of the whole; once its function is understood in relation to the whole, the cause is known.

In Chapters 5 and 6 we gave two examples of *non*-teleological explanations from the Scientific Revolution period – Toricelli's vacuum and Harvey's circulation of the blood – and noted that these were inherently progressive rather than 'final'. Aristotle's teleological explanations in biology were in this sense 'anti-progressive'. His account of the human hand is an example. The human hand fascinated the ancients, who wondered why it is so well constructed for the work it does. Aristotle's answer stood for more than 2,000 years: *humans have hands because they are humans and need them.*[4] That answer does not lead to further inquiry; it is 'final'. From a modern scientific standpoint, that is considered an objection to teleological explanations in general. They do not form a basis for experimental inquiry.

De Motu Animalium and De Incessu Animalium

Many of Aristotle's generalisations about animal locomotion seem commonplace to us, but they were entirely new when they were written and testify to a breathtaking range of observational data. Fish move from place to place only by swimming, most mammals only by walking or running. However, no animal can move *only* by flying; the ones that fly can also walk. Some animals move about and some do not; the ones that do not are *only* found in water. However, Aristotle went far beyond these generalisations.

Most impressive to modern readers was his grasp of what we now call *homology*. In *De Incessu* he wrote:

> Birds have wings in the upper part of their bodies and fishes have two fins in the front part of their bodies. Birds have feet on their under part and most fishes have a second pair of fins in their under part…

He had recognised that the fins of a fish and the wings and legs of a bird are *homologous structures*, different in appearance but nevertheless basically the same; an extraordinary inference given the absence of background knowledge. He made no explicit claim about evolution[5]; yet more than two millennia later, the existence of homologous structures among closely related groups of organisms was to become one of the principal lines of evidence in Darwin's *Origin of Species*.

---

[4] Galen, a devoted follower of Aristotle, thought the hand so important that he devoted the entire first chapter of his *Uses of the Parts* to this teleological explanation of its structure and function.

[5] This is a contentious point; Darwin, for example, asserted that Aristotle had grasped the idea of evolution, but others deny it. Certainly Aristotle was aware of a tradition of evolution-like ideas in pre-Classical Greece. He attributes to Empedocles a notion similar to natural selection: *'Wherever therefore all the parts came [to be] just what they would have been if they had come to be for an end, such things survived, being organized spontaneously in a fitting way; whereas those which grew otherwise perished and continue to perish…'* Centuries later, Islamic philosophers were to make similar observations (Chapter 4).

De Generatione Animalium

In the *Historia Animalium* we find a detailed *description* of chick embryo development (compare Fig. 7.1). By examining eggs on successive days of incubation, Aristotle identified the sequence in which the parts of the chick's body first appear. The account is flawless; it is only thanks to the microscope that we can now add many further details. In *De Generatione* we find a comprehensive attempt to *explain* that process of development. In effect, Aristotle now applies the principle of causation developed in *De Partibus* to the studies of what we would now call reproduction, development and heredity.

De Generatione contains some of the most difficult but rewarding passages in all of Aristotle's surviving work. We are reminded that the chick embryo is formed part by part, the heart before other internal organs, the head and eyes before the limbs. Moreover, *all* bird embryos follow the same developmental sequence. That is a remarkable inductive generalisation; once again, it must have come from a large body of observational evidence. But it is more than that; it refutes the 'preformationist' claim that the entire adult organism is present in miniature from the time of conception and merely *grows* during gestation. Instead, it supports the notion of 'epigenesis', which holds that biological forms develop from apparently formless beginnings. Thus, Aristotle resolved one of the great disputes in the history of biology, preformationism versus epigenesis, two millennia before that dispute actually began.

That alone would have been a striking achievement, but Aristotle went further. First, he observed that although some animals produce eggs that are incubated externally, others do not; yet they all develop through similar embryo stages. Therefore, eggs cannot be 'fundamental' to embryo formation. This led him to infer that the male and female parents contribute some 'substance' to the production of the new individual. He called this substance 'semen'; nowadays we would use the word 'gametes'. He reasoned that the 'semen' cannot be produced from (and cannot represent) the entire body. On the contrary, an entire new body is produced *from* it. In other words, he denied the notion of *pangenesis* (all parts of the body contribute to what is inherited by the offspring), which was to be proposed by Darwin and hotly debated in the late 19th century.

Secondly, he recognised that this chain of observation and reasoning pointed to a very profound conclusion about the nature of living things: they arise from matter that has the 'potential' to form them, but that form is 'actualised' by some impulse that exists within them. A seed contains the potential to make a new plant, but that potential may or may not be actualised. He coined a word to denote this process of actualising potential, of development towards a predetermined end: *entelechy*. Translated literally from its Greek roots, 'entelechy' means 'that which holds within itself the completion of the end'.

The view taken in *De Generatione* is that the female parent provides or contains the potential, which is actualised by the 'heat' of the male parent. Nomenclature aside, this account is not incompatible with our modern understanding of fertilisation and development. But Aristotle hesitated to generalise his pronouncement to

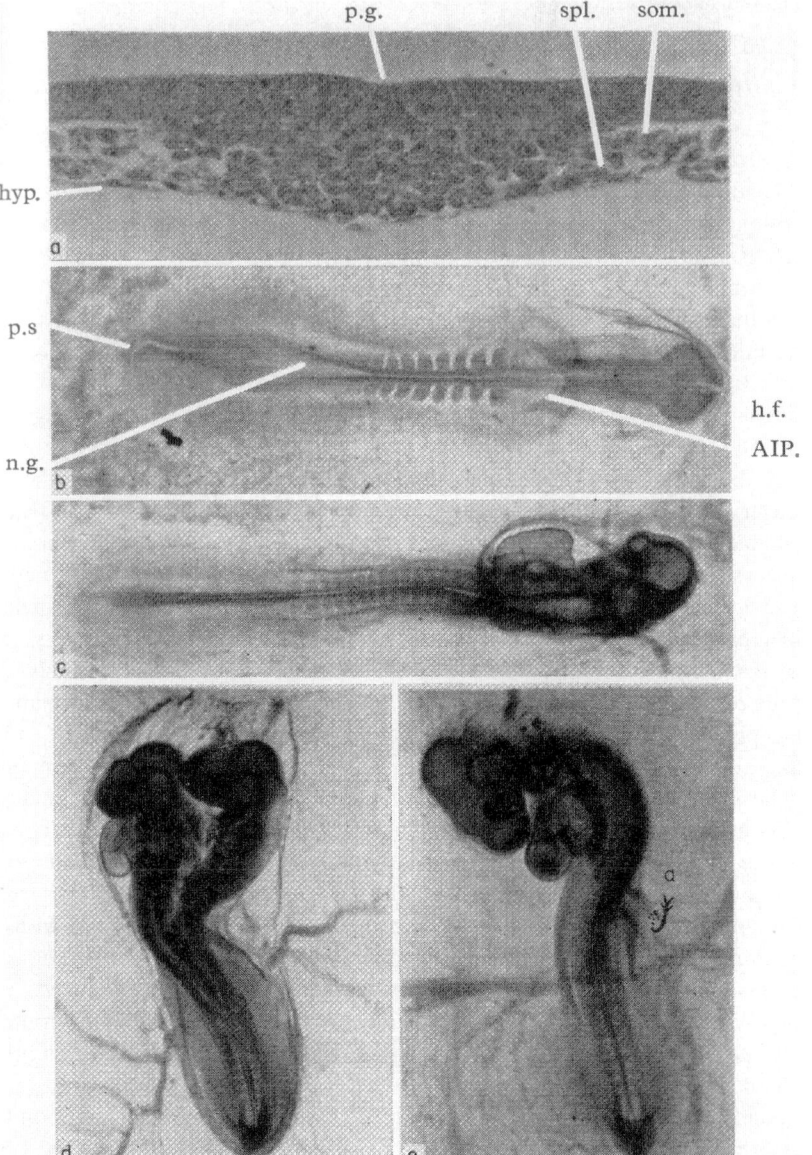

Fig. 7.1 Stages in the development of the chick embryo. Image (a) is a micrograph of a section through a very early embryo, showing structures that will develop later into adult tissues (key: p.g. = primitive groove, leading to opening in blastula; spl. = splanchnopleure and som. = somatopleure, two layers into which the mesoderm splits; hyp. = hypoblast, tissue forming from the inner cell mass). Aristotle would not have been able to see this detail, but clearly described the appearance of the 24-hour (b), 36-hour (c) and 72-hour (e) embryos. (d) is a dissection of the 36-hour embryo made by longitudinal cuts through the blastoderm. Key: p.s. = primitive streak (in which the primitive groove forms); AIP = intestinal portal blood vessel; n.g. = neural folds, early development of spine. The head and trunk are clearly visible at 36 hours (c) and the eyes and the beginning of the tail at 72 hours. The history of embryology is outlined in Chapter 10 and further details are given there. Reproduced with permission from Cohen J (1963) Living Embryos. Pergamon, London

all organisms. Some living things, he averred, arise from non-living matter without sex; in other words, he believed in *spontaneous generation* of some species.

Nowadays, we reject both spontaneous generation and the entelechy concept. Yet it is not immediately obvious how 'entelechy' differs from the modern idea that development is genetically programmed. In a sense, the fertilised egg has 'potential' and its genotype determines, and executes, the 'actualisation' of that potential. Perhaps it is not surprising that the debates arising from these Aristotelian notions continued for centuries after the Scientific Revolution.

De Anima

Aristotle's term ψυχη (psyche) was translated into Latin as *anima* (= soul), and this has led to misunderstandings. Aristotle *defined* 'psyche' as '*the first entelechy of an elementary natural body which potentially possesses life and which is the instrument of the soul*'. In other words, it is the 'actualiser of potential', that which realises the formal and final causes, and the body is its *instrument*. Perhaps a musical metaphor was intended; the body may be thought of as the instrument upon which the 'psyche' or 'anima' plays, producing the music of biological organisation. In Book I of *De Anima*, the 'soul' is described rather more simply as '*that by virtue of which living things have life*'. In short, it is the distinction between the living and the non-living.

On first acquaintance, this pronouncement may seem confusing. We recall that in *De Partibus*, Aristotle ascribed the 'livingness' of animals to *pneuma*, the vital spirit that is taken into the body by respiration. So what is the relationship between *psyche* and *pneuma*? Crucially, *psyche* is inherent in the organism whereas *pneuma* is supplied from the environment. *Pneuma* is required to make *psyche* operate, and *pneuma* works as it does because *psyche* directs it. Also, *pneuma* is the provider of vital spirit only for air-breathing animals, not for other animals or for plants,[6] whereas (by definition) all living organisms have a 'soul'.

In Book II of *De Anima*, Aristotle distinguishes three kinds of soul, or three facets of soul: nutritive, perceptual and rational. All organisms nourish themselves and reproduce, so they must have nutritive souls. Animals also have some form of sense perception ('animals with blood' have several senses) so they have perceptual as well as nutritive souls. Perception of any kind entails the ability to feel pleasure and pain, and that distinction in turn entails the existence of desire. In addition, some animals have the powers of self-motion, imagination and memory.

---

[6] Theophrastus (371–287 BC), another pupil of Plato, became master of the Lyceum after Aristotle died. Like his older colleague he was a man of wide learning, but none of his works have survived except for his *Inquiry into Plants*, an attempt to emulate Aristotle's *Historia Animalium* in the sphere of what we now call botany. Together, Aristotle and Theophrastus constructed a biology that survived until after the Scientific Revolution. We probably underestimate the latter scholar because most of his work has been lost.

Many commentators question whether *De Anima* should be counted among Aristotle's 'biological' works. Book III is definitely not part of Aristotelian biology (it is concerned with the rational nature of humans) but Books I and II contain ideas that are central to the concept of entelechy and therefore to the whole Aristotelian notion of biological causation, which were to underpin all the major debates in biology and medicine after the Scientific Revolution. We therefore count *De Anima* as 'biological'.

Aristotle, Harvey and Descartes

Aristotle devotes the whole of Book III of *De Anima* to the rational soul, which is unique to humans (the only rational animals). In contrast to the nutritive and perceptual souls, which are clearly biological ideas, the 'rational soul' is closer to our more familiar religious understanding of 'soul'. It is interesting that Descartes sought to eliminate Aristotle's notion of entelechy and thus the nutritive and perceptual souls, but retained the rational soul as the distinguishing feature of humans. Descartes believed that his 'mechanical' account of organisms obviated the need for entelechy, but he was obliged to equip humans with a soul in order to retain their status as God's special creation.

In several of his writings before the early 1630s, Descartes pronounced an unreservedly mechanistic account of the universe. His famous 'dualism' (human bodies are machines but their souls are divine) became apparent in the *Discourse on Method* of 1637. It is often said that after the trial and house arrest of Galileo in 1632–1633, Descartes developed a paranoid fear of reprisal from the Church and lived the rest of his life like a hunted man. That may be true, but his *Treatise on Man*, which was written *before* Galileo's arrest (though not published until later), is explicitly dualistic.[7] Along with his uncertain attempt to explain nerve function (Fig. 7.2) and the role of 'animal spirits' (Chapter 6), this shows how difficult it was for Descartes to break free of Aristotelianism in biology – in contrast to physics, where his stance was unreservedly anti-Aristotelian.

Harvey, however, was anti-Aristotelian in a methodological but not a conceptual sense. His natural-philosophical approach to study (meticulous experimenting, observing and reasoning) destroyed the Galenist, and therefore the Aristotelian, account of heart function and blood movement. However, the language in which he described the circulation remained Aristotelian *in character*, though the details were new. This is most apparent in his attribution of 'life-giving qualities' (or 'life-bearing qualities') to the blood. Harvey suppresses any discussion of a 'soul', which is not observable or measurable, but it is clear that he saw the self-moving, circulating blood of animals as both the manifestation and the *modus operandi* of

---

[7] First published in 1664. Translated Hall TS (1972) Harvard University Press, Cambridge MA.

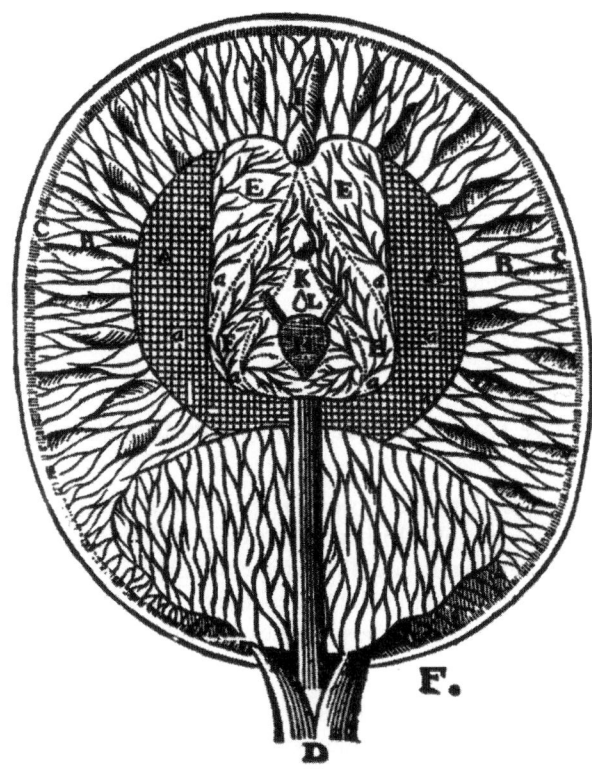

Fig. 7.2 The control of muscle contraction by the brain as envisaged by Descartes. According to this model, which was soon rejected by Descartes's successors, animal spirits are secreted from the pineal gland (H) into the ventricles (E) and make their way through pores (a) in the ventricle walls (A) into tubes (C) that pass into the spinal cord (D) and from there to the muscle. This input causes the muscle to swell, leading to muscle contraction. When the inclination of the pineal gland changes, the rate of inflow of animal spirits into this hydraulic system is altered. Reproduced from Fig. 12.4 (p. 98) in Lutz PL (2002) The Rise of Experimental Biology. Humana, New York. With kind permission of Springer Science and Business Media

entelechy.[8] Harvey thus avoided the dualism that Descartes was compelled to adopt; by grafting Aristotelian explanations on to natural-philosophical methods and observations, he retained Aristotle's monism.

   Much of Harvey's work appears to have been lost during the Civil War, the Commonwealth and (nine years after his death) the Great Fire of London, but his surviving writings testify to this 'conceptual Aristotelianism'. He could be critical of Aristotle; he was always prepared to challenge The Philosopher on matters of fact. But he never seems to have doubted Aristotle's approach to biological

---

[8] More than 100 years later, a distinguished follower of Harvey, the great surgeon John Hunter, remarked that 'blood has life'. Many commentators have been puzzled by this comment, but Hunter was only echoing Harvey – and, in spirit, Aristotle.

causality. The most notable example was written when he was 73 years old, in 1651, after the king whom Harvey had served as court physician had been executed and his political enemies had taken power. The title is deliberately Aristotelian: *Essays on the Generation of Animals*. This book contains the famous quotation '… *omnia omnino animalia, etiam vivipara, atque hominem adeo ipsum, ex ovo progigni*'[9] – later abbreviated to '*omne vivum ex ovo*', all living things come from an egg.[10] Although more precise than Aristotle's vague 'semen' concept, the idea remains Aristotelian at heart: eggs are objects with potential, which can be actualised by their intrinsic entelechy to yield final forms. Yet at the same time, Harvey appears to have rejected another Aristotelian notion, spontaneous generation.

Biology after the Late 17th Century: Aristotle's Legacy

Within a generation of Harvey's *Essays on the Generation of Animals*, one early microscopist (van Leeuwenhoek) had proposed that fertilisation of an egg by a spermatozoon was universal in sexual reproduction, while another (Swammerdam) had also rejected the idea of spontaneous generation (see Chapter 6). Descartes and Harvey represented two approaches to the nascent 'sciences' of biology and medicine:

1. Attempts to overturn Aristotle's biology as his physics had been overturned, by accounting for living structures and functions in terms of mechanics. Because Descartes took this approach (not altogether successfully or consistently), and because Descartes was pre-eminently a creator of conceptual systems, many of his successors during the next two or three generations also created systems of mechanical explanation in medicine and biology.
2. Attempts to interpret natural-philosophical discoveries about organisms in terms of the Aristotelian model of causation, and particularly by reference to entelechy. Paradoxically, many who adopted this approach in the late 17th and 18th centuries were, like Harvey, methodologically Baconian; they emphasised observable facts and proceeded cautiously towards inference and generalisation. For many of them, such as the great physician Thomas Sydenham (1624–1689, known as 'the English Hippocrates'), the Aristotelian viewpoint remained in the background but became apparent, for example, in references to the 'wisdom of Nature'.

---

[9] '… absolutely all animals, including those who give birth to live young, and even man himself, are born from an egg'.

[10] This was a remarkably prescient idea: the mammalian egg was not discovered until 1827, by the Estonian biologist Karl von Baer (1792–1876). Regnier de Graaf (1641–1673), influenced by Harvey, described follicles that he believed were mammalian eggs, but von Baer later showed that these follicles were not ova; they *contained* the ova. Today they are known as Graafian follicles.

Chapter 8
How Different Are Organisms from Inanimate Objects?

The conceptual schism in biology and medicine outlined at the end of Chapter 7 can be construed as the beginning of the 'mechanism-vitalism' debate, but that is simplistic and rather misleading. 'Vitalism' is a slippery notion at best, and a number of recent commentators have applied the word inappropriately, even to such work as Harvey's.[1] The word 'mechanism', too, denotes several mutually incompatible variants of the Cartesian position. The following examples date from 1695 to 1747.

Organisms as Mechanisms

After Descartes, the application of mechanics to the human body was pioneered by Giovanni Borelli (1608–1679). His posthumous *On the Motion of Animals* (1680) is widely recognized as a seminal work. A generation later, his example was followed by Friedrich Hoffmann (1660–1742), an ideal Cartesian, a committed mechanist and an inveterate systematiser. He held that bodies must be considered as machines composed of fluid and solid parts. Their workings in both health and disease must be understood in terms of the motions of particles, which are matters of mechanics and hydrodynamics. Different kinds of particles differ in chemical properties, but these too must be conceived in mechanical terms. Chemical attraction was believed to be mechanical.[2]

---

[1] 'Vitalism' is the belief that living organisms are distinguished by a 'force' or 'spirit' that disappears when they die and has no counterpart in the non-living world. 'Mechanism' is the belief that organisms and non-living objects are subject to exactly the same laws (of physics and chemistry). It is popularly supposed that biology was once dominated by opposition between proponents of these two beliefs, but mechanism finally won the battle because it was scientific while vitalism was mystical. That is a caricature of history.

[2] This Newtonian idea persisted well into the 19th century. It was not until the later part of that century that the nature of chemical bonds came to be understood in something like the way we understand it today.

Was Hoffmann any more able than Descartes to sustain this radically anti-Aristotelian stance? His *Fundamenta Medicinae* of 1695 opens with the Cartesian pronouncement:

> Medicine is the task of utilising physico-mechanical principles properly, in order to conserve the health of man or to restore it if lost.

However, in the sixth chapter of Book I we find the following:

> The animal spirits, provided with the utmost fineness and elasticity, have a power impressed by God, not only of moving themselves mechanically, but doing so by choice, purposefully and towards a definite goal.

In other words, Hoffmann could sustain his mechanical view of physiology and medicine only at the cost of re-introducing Aristotle's entelechy, and by invoking God as a causal agency in a natural process, contrary to the spirit of the Scientific Revolution.

Herman Boerhaave (1668–1738), the great medical teacher of Leiden University, offered a mechanistic system that was more complete and much more influential than Hoffmann's. He integrated recent discoveries in mechanics and chemistry into a unified whole, balancing observation with reason as Galileo and Harvey had done. Boerhaave repudiated any distinction between the material components of organisms and inanimate objects. He wrote:

> The universal laws of nature… depend on mechanical and physical principles… the same laws are also true in the human body, for its matter appears to be universally the same with that of all other bodies'.[3]

Elsewhere in the same work we find:

> Mechanics… supposes a previous knowledge of the structure of all the parts in the human body, to which would apply mechanical laws; and in this sense physic is no more than the knowledge of such things as are transacted in the human body.

There seems to be no covert Aristotelianism here, as there is in Hoffmann's work. However, we also find:

> Primary causes are those productive of secondary ones; but we always meet with God in our search after these, and this puts a stop to our further knowledge.

Thus, at some point in our search for mechanistic causes in biology and medicine, we are obliged to have recourse to God. Aristotelianism cannot be avoided in any other way.

Julien Offray de la Mettrie (1709–1751) studied medicine under Boerhaave at Leiden. He was neither a clever experimenter nor a systematiser but he was a lucid, witty and satirical writer. Literate and well-informed, he greatly admired Descartes and Boerhaave but wanted to show that mind (= soul) is a product of body, that the universe contains only one kind of substance and that there is no need to invoke God as a causal agency. His *Histoire Naturel de l'Ame* (1745) was a satirical attack

---

[3] Academical Lectures on the Theory of Physic, vol. 1 (1741).

on Aristotelian thought and on religion, and was duly denounced for its atheistic materialism. *L'Homme Machine* (1747) propounded even more extreme views. It was mischievously dedicated to Albrech von Haller, a fellow-student of Boerhaave known for his piety, and La Mettrie's books were thereafter burned even in liberal Holland. After 1748 La Mettrie fled to the court of Frederick the Great of Prussia and developed implicitly 'evolutionary' ideas.

At the end of *L'Homme Machine* he wrote:

> Let us then conclude boldly that man is a machine, and that in the whole universe there is but a single substance differently modified. This is no hypothesis set forth by dint of a number of postulates and assumptions; it is not the work of prejudice, nor even of my reason alone; I should have disdained a guide that I think so untrustworthy, had not my senses, bearing a torch, so to speak, induced me to follow reason by lighting the way themselves. Experience has thus spoken to me on behalf of reason; and in this way I have combined the two.

What had happened to European thought during the century following Harvey and Descartes that allowed religious considerations to be sidelined, and even dismissed, in discussions of biology and medicine, and Aristotle to be apparently so despised? How, in other words, can we account for the transmutation in mechanistic thought illustrated through the writings of Hoffmann, Boerhaave and la Mettrie?

Locke: Classical Empiricism

John Locke (1632–1700) was one of Newton's few personal friends. His *Essay Concerning Human Understanding* established *classical empiricism*, which elaborated the philosophies of Bacon and Gassendi. Locke believed that the mind at birth is a blank slate and only the experience of the senses can write on that slate, generating 'ideas'. Thus, all true knowledge is rooted in observation. We arrive at understanding by induction, i.e. we forge connections between distinct ideas because those ideas regularly occur in association or sequence. (The phrase 'association of ideas' is Locke's.) Locke also accepted atomism, which had become the consensus view. He was very influential because his writing style was simple and lucid and because he was closely acquainted with Newton, the ultimate exemplar of natural philosophy. He spoke for the whole new way of thinking. In chapter 2 he wrote:

> Though the qualities that affect our senses are, in the things themselves, so united and blended, that there is no separation, no distance between them; yet it is plain, the ideas they produce in the mind enter by the senses simple and unmixed.

Part of the motivation for Locke's *Essay* was to convince the world that Newton's achievements had resulted from strict application of the Baconian precepts, beginning afresh from observation and avoiding all preconceptions. In particular, the laws of classical mechanics were powerful and universal because they depended on the most basic, least 'idol-strewn' of all languages: mathematics. The founders of the Royal Society declared that the proper language for seeking true knowledge should be as simple, direct and 'mathematical' as possible. Their opinion came to permeate western European thought. During the late 17th and 18th centuries, essays, poetry, church sermons, political speeches and indeed all forms of written language were shorn of the rhetorical flights of earlier ages and brought closer to the Lockean ideal. The joint influences of Locke and Newton spread through all aspects of culture. Both Boerhaave and la Mettrie were unstinting in their admiration for Locke.

The Enlightenment

Unlike Bacon, Locke applied his empiricist principles to all walks of life. Just as there is natural philosophy, a fundamentally correct way of understanding the natural world that was exemplified by the work of Newton, so there are natural morality, natural justice, natural religion and so on. This view underpinned the most radical developments of European thought and politics during the 18th century. In Paris, Leiden, Edinburgh and other cities, it gave rise to a body of ideas and writings subsequently known as *The Enlightenment*. The 18th century has been called the 'Age of Reason' because of this cultural shift.

One effect, or symptom, of the Enlightenment was to make studies of living organisms in their natural habitats and in captivity, of rocks and landform, of chemical processes, of the worlds made visible by microscopes and telescopes, into topics of popular discourse among the gentry. Such studies became the pursuits or pastimes of wealthy amateurs. They were described in the *Philosophical Transactions of the Royal Society* and in numerous other publications. In terms of scientific content and intrinsic interest, Gilbert White's *Natural History of Selborne* is an outstanding example of this trend, but in spirit it typifies the age. By 1800 the names of now-familiar scientific disciplines were coming into use: geology, chemistry, biology.

Natural or experimental philosophy rose to intellectual pre-eminence within a couple of generations, celebrating new knowledge and discovery. Its supposed pioneers (notably Newton and Locke) were lauded by most thinking people and vilified by a minority; no one was indifferent to their influence.[4] Concomitantly, Plato and Aristotle became targets of satire, not least from the pen of Voltaire,[5] the

---

[4] Locke was just as metaphysical as Descartes, and Descartes just as 'scientific' as Galileo and Newton. All these men agreed that experiment and observation were the irreducible basis for reliable knowledge about the world, including organisms. By the early 18th century, this belief had attained widespread official blessing and encouragement. Few people questioned the stance of the natural philosophers, and those few were mostly confined to doctrinal institutions. Patronage, often extended by Church princes, wrought a great deal of influence on 18th century thought, and various societies arose in which 'empiricism' was deliberately cultivated.

[5] These attitudes are explicit in Voltaire's *Letters on England*. After the collapse of Soviet Communism, Russian satirists publicly lampooned Marx and Lenin; the lampooning of Plato and Aristotle during the Enlightenment, notably by Voltaire, was analogous.

corrosive cutting edge of Enlightenment thought. The need to replace Aristotelianism in biology with a 'Newtonian' account of organisms became urgent, but the problems of entelechy and 'animal spirits' were reluctant to yield to the demands of fashion.

Concurrently, the spirit of the Enlightenment entered other aspects of life. European nations were inspired with new confidence in their capacity to make the world a better place. They could boost their own prosperity and wellbeing by following Baconian precepts and could export their emerging recipe for Earthly Paradise to the rest of the globe. This was the beginning of the great era of colonial expansion. Political debate took on new directions; from Locke's writings emerged the concepts of freedom, equality and natural human rights, radical notions that were to become the roots of modern democracy. The long process of secularisation had at last found full articulation.

Late in the 18th century the combination of social and intellectual change, radical political ideas and increasing secularisation culminated in the revolt of the American colonies and in the French Revolution. The American Declaration of Independence quotes wholesale from Locke; significantly, its two principal authors, Thomas Jefferson and Benjamin Franklin, were considerable natural philosophers. Within a hundred years of its inception, the empiricist tradition articulated by Locke had wrought radical changes not only in the way in which western Europeans thought about themselves and the world in which they lived, and how they could best exploit the world for their own benefit, but also in the ways in which they chose to organise it politically.

Against Mechanism

The hope that all determinable aspects of the world would be explicated mathematically – parts, motion, time, space – had faded by the end of the 17th century. Had it not been for the invention of the calculus, the 'mechanistic' ideal of natural philosophy might have died in the cradle. In the event, however, Lockean ideas spread through France and then through the remainder of Europe during the first half of the 18th century; 'mechanism' took on a different and more specific meaning, which derived from Descartes's view of animals as machines rather than 'magic'. This sense of 'mechanism' evolved through the writings of Hoffmann, Boerhaave and la Mettrie (see above) as the Enlightenment advanced.

The difficulty, which not even la Mettrie could circumvent, was that mechanism failed to account for the *obvious* differences between living and non-living, particularly the *purposiveness* of organisms and their parts. Mechanistic natural philosophy had rid the inanimate world of purpose. Mechanists in medicine and biology faced the seemingly impossible task of explicating the animate world, in which purpose is inherent, in terms that repudiate purpose. The mechanistic account of biology therefore had critics and detractors, who broadly adopted one or more of the following three standpoints:

1. Organisms and inanimate objects consist of different kinds of matter, contrary to what Boerhaave and la Mettrie said. In Aristotelian language: living (organic) matter has potential that can be actualised, but non-living (inorganic) matter does not. This was contrary to the Cartesian and Newtonian assumption that matter is everywhere the same.
2. Organisms and inanimate objects consist of the same kind of matter and are subject to the same kinds of forces, but there is an additional force in organisms that accounts for their distinctively 'living' properties such as purposiveness. In Aristotelian terms: this force is responsible for actualising the potential of matter in organisms (but should it be considered to correspond to *pneuma* or to entelechy?).
3. Organisms and inanimate objects consist of the same kind of matter and are subject to all and *only* the same kinds of forces, exactly as the mechanists say. However, living matter differs in the way in which it is *organised*, and this organisation leads to the distinctive properties of living entities. In Aristotelian terms: entelechy resides in the organisation of matter.[6]

Strictly speaking, 'vitalism' denotes only the second of these three positions, but it has been loosely applied to all of them. Such loose usage has effectively emptied the word of meaning. Nowadays, unless its use is specified precisely, 'vitalism' can only be taken to indicate a point of view or an assertion about biology that is scientifically and philosophically unacceptable. The user is merely saying 'I reject this opinion'. 'Vitalism' has become a pejorative term with no clear or single meaning, more an expletive than an abstract noun.

Living and Non-Living Matter; 'Vital Force'

We can now explore some developments of these anti-mechanist positions during the 18th and 19th centuries.

The Scientific Revolution coincided with the first significant advances in chemistry since Roger Bacon's reiteration of the studies of Geber and his Muslim successors. Theories of alchemy entailed an *organic* conception of the matter: to change one substance into another, the spirit responsible for its properties must be replaced. Every different substance comprised a particular combination of the four elements of Empedocles (earth, air, fire and water). Alchemists were generally secretive about their work, which involved magic. Such was the background against which a natural-philosophical approach to chemistry began slowly to emerge.

Santorio Santorio (1561–1636) was one of the first to apply the philosophy of the Scientific Revolution to animal biology. His experiments laid the foundation for the study of metabolism and the physical and chemical processes of the human

---

[6]This, in effect, is our modern view in biology (*not* usually dismissed as vitalistic!) but it is also the stance adopted by Harvey (which *has* been dismissed as vitalistic; Chapter 6) and Hunter ('*a dead body has all the composition of a living one*' – also deemed 'vitalistic' by some writers).

body. But it was shortly to be superseded by the highly influential work of Jean-Baptiste van Helmont (1579–1644), which combined the late mediaeval alchemical and magical traditions with the new mechanistic ideas.

Helmont

Helmont recognised that there are gases other than air (carbon dioxide, carbon monoxide, nitrous oxide and methane) and allegedly coined the word 'gas' (from the Greek χαος = 'chaos'). He devised a temperature scale based on the melting point of ice and the boiling point of water. In medicine, he used remedies that specifically considered the type of disease, the organ affected and the causative agent.[7] He showed that the digestion of food in the stomach involves the production of acid. By carefully weighing all his materials before and after his experiments, he found that matter, including the food we eat, cannot be destroyed but only changed in form. He was critical of the traditional 'four elements', rejecting fire as an element and reasoning that earth is really only water. His evidence for that was to grow a willow tree in a measured quantity of earth, adding only water. Over a period of five years the tree became heavier, i.e. matter was added to it, but the weight of the earth in which it grew was unchanged; therefore, the matter of the tree could only have been formed from water and air. Importantly, Helmont said that the behaviour of living matter is distinctive because it depends on an *archaeus*, which corresponds approximately to Aristotle's 'nutritive soul'.

Sylvius

Franciscus Sylvius (1614–1672) introduced the idea of 'chemical affinity' to explain the human body's use of salts. He and his followers made significant contributions to the study of digestion and of body fluids.

Boyle

In *The Sceptical Chymist*, Robert Boyle insisted that matter be studied by experiment rather than by speculation, an echo of the teachings of Geber and of Roger Bacon,

---

[7] He claimed to be a follower, albeit a critical one, of the eccentric Theophrastus Philippus Aureolus Bombastus von Hohenheim (1493–1541), better known as Paracelsus. Paracelsus was a practitioner of astrology, alchemy and magic but reacted strongly against standard mediaeval teachings, attacking Galenism and publicly burning a copy of Avicenna's *Canon of Medicine*. Paracelsus used mineral cures including mercury to treat the sick, often with success, and he apparently discovered and named the element zinc, but his abrasive and uncompromising behaviour earned him far more enemies than friends.

and that the results of experiments must be made public, contrary to the secretive alchemical tradition. Like Helmont, he struggled with the traditional notion of 'elements' and found that it made no coherent sense. He gave the first description of acid-base indicators. A later work,[8] *Memoirs for the Natural History of Humane Blood*, was the first book dedicated to 'animal chemistry', as the subject came to be called. Written a generation after the death of Harvey, the *Memoirs* attempted to show that blood is as amenable to chemical analysis as any other substance. Implicitly, it suggested that matter obtained from organisms differs from non-living matter: the germ of the idea that 'organic' and 'inorganic' substances were different in kind.

This difference was 'obvious' to most investigators. For example, when organic matter was heated, even gently, it underwent dramatic changes; but when inorganic matter was heated, it remained unchanged or little changed.

Stahl

Georg Ernst Stahl (1660–1734), an exact contemporary and later a rival of Hoffmann, established that the corrosion (oxidation) of metals was essentially the same process as combustion. To explain the process he adopted[9] the *phlogiston* hypothesis. 'Phlogiston' was something that is lost during combustion (or corrosion), but Stahl also used the word to indicate 'the property of being combustible'. The former definition is mechanistic and accords with the spirit of the Scientific Revolution; the latter is Aristotelian in character.

Stahl was explicitly Aristotelian in his belief that living and non-living matter are different. He said that the behaviour of living matter did not depend exclusively on the motions of particles, as Hoffmann and Boerhaave would have us believe, but on the *anima*, as Aristotle maintained.[10] The *anima* integrates the individual movements of particles of matter into a well-ordered, goal-directed whole. Stahl pointed out that without recourse to a similar concept, mechanists such as Hoffmann and Boerhaave could not explain the integrated nature of organisms or their purposiveness.

It was hard to gainsay that point. Stahl's critique of Hoffmann's extreme mechanism was penetrating and it greatly influenced 18th century thought (as did his version of the phlogiston hypothesis). It suggested that the difference between organic and inorganic matter might not lie in the existence or otherwise of a potential to be actualised, but in the intrinsic presence of an agency that could actualise it. To an extent, this conflated the first and second anti-mechanist positions (see above).

---

[8] Published in 1684; the first edition of *The Sceptical Chymist* dates from 1661.

[9] The phlogiston hypothesis was first proposed by the Johann Joachim Becher (1635–1682), and may have been suggested by van Helmont, but Stahl developed it into an explanatory notion that survived until 1790. Some writers have described this hypothesis as 'vitalistic'. Since it relates to chemistry in general, not specifically to biology and medicine, it is difficult to know what they mean by that – apart from 'wrong'.

[10] A number of commentators have equated Stahl's use of *anima* with Helmont's *archaeus*.

Buffon

George-Louis Leclerc, Comte de Buffon (1707–1788), was the doyen of French naturalists. His multi-volume *Histoire naturelle, générale et particulière, avec la description du cabinet du roi* (1749–1789), an encyclopaedia of biological knowledge, is accounted a literary masterpiece. To argue that there was no entelechy or intrinsic 'finality' in organisms, he was obliged to attribute new powers to matter to account for vital action; this was the first clear and explicit distinction between organic and inorganic substances. Many of his ideas seem speculative, but Buffon was highly influential and he was a committed empiricist in the Bacon-Locke tradition. He believed that inductive reasoning from consistent observations led to greater certainty than did abstract mathematical and deductive argument.

Casimir Medicus

Animal chemistry advanced during the 1760s and 1770s when physicians began to apply chemical principles to the studies of digestion and respiration as well as the composition of the blood. It soon became clear that living processes involved complex chemical transformations that could not be replicated in the laboratory, so Stahl's conclusions seemed to be supported. Accordingly, Friedrich Casimir Medicus (1736–1808), a celebrated botanist, introduced the term *Lebenskraft*, or 'vital force', to animal chemistry in 1774. The vital force 'explained' those chemical processes in organisms that apparently could not be explained in terms of the chemistry of inanimate matter. Medicus seems to have intended *Lebenskraft* to be understood as a 'force' in something akin to the Newtonian sense: he propounded the second, not the first, anti-mechanist position.

Lavoisier

Antoine Lavoisier (1743–1794) developed the law of conservation of matter, implicit in Helmont's work, and expounded it as a firm quantitative statement. His work was instrumental in the emergence of scientific chemistry. Famously, he showed that combustion, respiration, corrosion and the formation of acid all required a particular component of the air, which Joseph Priestley had called 'dephlogisticated air'. Lavoisier renamed this component 'pure air' and subsequently *oxygen* (= acid producer). This work refuted Stahl's phlogiston hypothesis. Less famously, but no less importantly, it overturned a key Aristotelian idea that had been adopted by Descartes and Hoffmann as well as by Harvey: what we obtain from the air we breathe is not 'animal spirit' or *pneuma* but a definite chemical substance, oxygen.

Moreover, Lavoisier realised that animal heat is produced by respiration, i.e. during the apparent[11] conversion ('decomposition') of pure air (oxygen) to fixed air (carbon dioxide). He wrote:

> Pure air, in passing through the lungs, undergoes then a decomposition analogous to that which takes place in the combustion of charcoal. Now in the combustion of charcoal the matter of fire is evolved, whence the matter of fire should likewise be evolved in the lungs in the interval between inhalation and exhalation, and it is this matter of fire without doubt which, distributed with the blood throughout the animal economy, maintains a constant heat… only those animals in nature which respire habitually are warm-blooded and … their warmth is the greater as respiration is more frequent; that is to say, there is a constant relation between the warmth of an animal and the quantity of air entering, or at least converted into fixed air in, its lungs.

Dalton and Berzelius

The advances in chemistry during the later decades of the 18th century culminated in the work of John Dalton (1766–1844). Dalton gave us the conceptions of 'element' and 'atom' that are familiar to us today: all matter is made of atoms; all atoms of a given element are identical; every chemical reaction is a rearrangement of atoms; compounds consist of molecules, which are specific combinations of atoms.

Nothing in Dalton's work suggested a distinction between organic and inorganic matter, but that issue remained unresolved. The inventor of the now-familiar symbols for the chemical elements, Jöns Jakob Berzelius (1779–1848), rejected the idea of a vital force but nevertheless argued that a 'regulative force' must be inherent in living matter to maintain and integrate its functions. This is close to the third antimechanist stance, that the difference between the living and the non-living lies in organisation (see below).

Wöhler

Friedrich Wöhler (1800–1882), a student of Berzelius, isolated a number of elements including silicon, aluminium and beryllium, and in collaboration with Liebig he made several advances in organic chemistry. He is most famous for synthesising an *organic* compound, urea, from an *inorganic* one, ammonium cyanate, in 1828, thus proving that chemical constituents of organisms can be made in the laboratory without the involvement of a 'vital force'.

---

[11] The oxygen atoms in the carbon dioxide we exhale do not, in fact, come from the oxygen we breathe in, but not until 20th century biochemistry was well advanced did that become known. The quoted passage is from the end of Lavoisier (1777) Memoir on combustion in general; Mémoires de l'Académie Royale des Sciences. From Leicester HM, Klickstein HS (1952), pp. 592–600.

Some popular histories have declared that Wöhler's synthesis of urea 'killed vitalism'. In view of the prominent scientists who *described* themselves as 'vitalists' after 1828, that cannot be true. What Wöhler really did was to show that there is no *fundamental* difference between organic and inorganic substances: both are subject to the same principles, the same laws of chemistry. Thus, neither the first nor the second anti-mechanist position is ultimately tenable. This inference was implicit in the work of Lavoisier and his successors, but Wöhler demonstrated it directly.

However, 'no fundamental difference' does not mean 'no difference at all'.

Pasteur

Among the many scientific achievements of Louis Pasteur (1822–1895) was the discovery that most molecules found in organisms have 'handedness'. That is to say, they can exist in at least two geometrical forms, each a mirror image of the other, but only one of these forms is normally present in organisms. Few inorganic chemicals have that property. During the early 1850s, Pasteur conducted a number of experiments in which he attempted to impose 'handedness' on synthetic organic molecules by, for example, exposing them to magnetic fields. These attempts failed, but it was clear that Pasteur recognised a real difference in chemistry between organisms and inanimate nature, and that he considered the possibility of converting the latter to the former.

Modern Beliefs about Living and Non-Living Matter

What has been the 'final' outcome of this history? As the early 19th century chemists showed, the first and second anti-mechanist positions have been refuted; the laws of chemistry apply equally to living and non-living matter. Nevertheless, living and non-living objects *are* chemically different. For example:

1. A large percentage of biological matter consists of the elements carbon, hydrogen, oxygen and nitrogen. In non-living matter such as clay and sand there is far less carbon and scarcely any nitrogen. On the other hand, such matter is rich in silicon and aluminium, elements that are rare in biology.
2. Most biological molecules are very much bigger than more complicated than those of non-living matter. The inorganic world has nothing remotely comparable to DNA or proteins.
3. Biological molecules have very specific shapes – the 'handedness' discovered by Pasteur – but non-biological molecules generally do not.

These differences do not imply a radical distinction between the living and the non-living, as vitalism would require, but they refute the extreme mechanistic pronouncement that all matter is 'the same'. How might Aristotle have reacted to our modern knowledge of biological chemistry? Would he have considered that entelechy resides in these real but non-fundamental differences between the molecules of life and those of the inanimate world? Perhaps he would have aligned himself with the third anti-materialist position, holding that entelechy resides not in the matter itself but in the way in which that matter is organised.

The Third Anti-Materialist Stance

The pioneering physiologist Johann Reil (1759–1813), attacking the concept of 'vital force', wrote:

> The whole body consists of several large components; each component again [consists] of muscles, vessels, nerves… Here is only the whole a machine and the parts of the whole are natural bodies without purposeful development. … Through the union of these countless organs, which by different stages combine together into a whole machine there are equally composite forces communicated to it.

On the basis of this account, which pictures the body as a hierarchically ordered machine obeying only mechanical principles, Reil has been accounted a 'vitalist' and a defender of the 'vital force' concept! That shows how absurdly the term 'vitalist' is sometimes used.

Several of the French pioneers of physiology during the early 19th century distanced themselves from the radical mechanism of Boerhaave and la Mettrie, but they seldom evinced belief in 'vital force'. Among this group, the most extreme anti-mechanist views were expressed by Marie-François-Xavier Bichat (1771–1802), who insisted that physiology be treated as entirely distinct from physics. At the beginning of his *Anatomie Générale* of 1801, published posthumously by his friends, Bichat wrote:

> In nature there are two classes of being, two classes of properties, two classes of science. Beings are either organic or inorganic; the properties, living or non-living; the sciences, physiological or physical… Sensibility and contractility are vital properties. Gravity, affinity, elasticity and the like, represent the non-vital properties.

He did not claim that organisms flout the laws of physics and chemistry, but rather that the laws (and methods) of physics and chemistry are not sufficient for our understanding. He insisted on careful and detailed analysis of living matter, relying on experiment and observation, and considered that the basis of pathology lies not merely in particular organs but in the tissues composing those organs. He described many differences among tissues and has been dubbed 'the father of histology'.

On the subject of 'sensitivity and contractility' Bichat remarked:

> The properties just analysed are not really inherent in the molecules of matter, but rather disappear as soon as the separate molecules have lost their organic arrangement. The properties belong exclusively to this arrangement…

It seems unhelpful and even misleading to describe Bichat as a 'vitalist'. He used the adjective 'vital' only to denote those properties of organisms that distinguish

them from the inanimate world (sensitivity and contractility). He did not presume
a hypothetical '*vital cause*' of such properties. His approach to physiology was
similar to Reil's: the distinctiveness of organisms lies neither in the matter from
which they are made nor in any mystical 'vital force', but in the organisation of
their components. This view has been echoed repeatedly during the two centuries
since Bichat and it is more or less what we believe today.

Nevertheless, Bichat's 'vital properties' relate to the *purposive* behaviour of
organisms: the capacity to detect relevant stimuli (sensitivity) and respond to them
in a way conducive to survival (contractility). Thus, what Bichat saw as distinctive
in organisms was Aristotle's entelechy by another name. Sadly, he did not live long
enough to articulate his views on embryo development.

Another eminent physician who discussed 'vital properties' as observable *effects*
or *consequences* of the living state, as opposed to mystical 'causes' of life, was
Joseph Lister (1827–1912). On the basis of a number of experimental studies,
Lister[12] concluded that '*loss or lack of vital properties*' was the common cause of
blood coagulation both in injured blood vessels and outside the body. He stated
explicitly that he did not believe in 'vital forces' but insisted, echoing Hunter, Reil
and Bichat, that '*living matter has properties distinct from, albeit not inconsistent
with, those that are traditionally addressed by physicists*'. Lister too has been
labelled a 'vitalist' (Fig. 8.1).

Overview

Not until more than a century had passed after the deaths of Descartes and Harvey
was the opposition between 'mechanism' and 'vitalism' in biology fully articulated. By then, several variants of Cartesian mechanism had been propounded
(Hoffmann, Boerhaave, la Mettrie) and at least three types of 'anti-mechanistic'
thought had emerged in biology and medicine. One of these, due in its explicit form
to Friedrich Casimir Medicus, *was* vitalism, i.e. belief in a 'vital force' peculiar to
organisms. The others held that either the nature or the organisation of living matter
made organisms distinctive. Both have merit in the light of modern knowledge.

Mechanists faced the intractable problem of accounting for the *purposiveness* of
organisms and their parts in terms of theories (those of physics and chemistry) that
do not admit of purpose. To evade this problem they either sneaked an extra 'vital'
element into their arguments, or ascribed purposes to God and therefore beyond the
purview of science, or ignored the problem as insignificant. None of these alternatives was, or is, satisfactory.

Anti-mechanists faced the equally intractable problem of reconciling their views
with a non-mystical approach to science. It was established that living and non-living

---

[12] This quotation is taken from Lister JL (1863) Croonian lecture: on the coagulation of the blood.
Proc Roy Soc Med 13:355–364, in which Lister discussed the coagulation of blood.

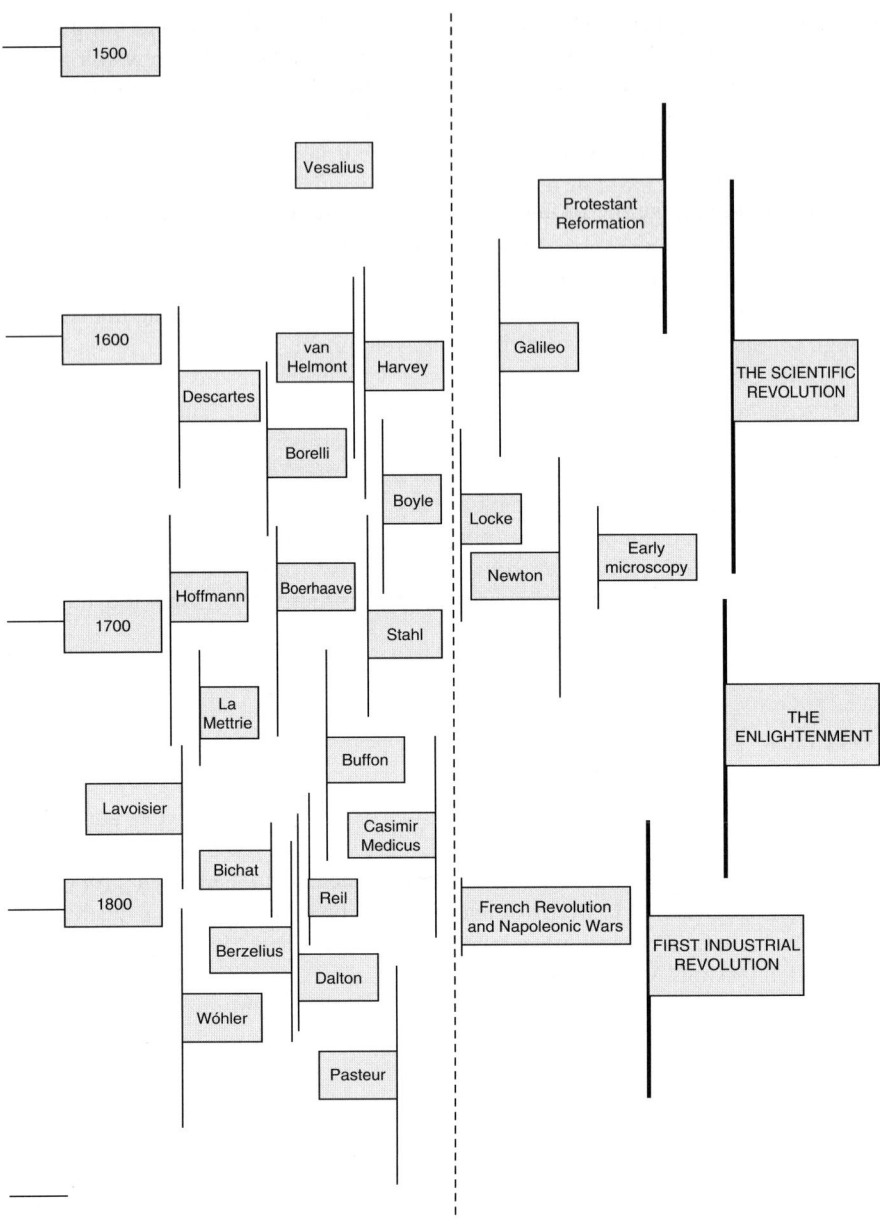

Fig. 8.1 Time-course of some salient contributions to thought and of events related to the early development of biology during the period 1500–1900, highlighting the dates of the Scientific Revolution, the Enlightenment and the first Industrial Revolution. This diagram may be seen as a continuation of Fig. 5.1 and may also be considered in conjunction with Fig. 9.1

matter do not differ fundamentally. The 'vital force' hypothesis was not only mystical but led to a viciously circular argument.[13] The uncontentious and indeed incontestable idea that 'vital properties' are consequences of the *organisation* of biological matter begs crucial questions: what is the cause of that organisation, and how does it lead to the 'vital properties' that entail purposiveness?

To sum up: neither mechanists nor anti-mechanists in the 18th and early 19th centuries could resolve the problem presented by Aristotle's entelechy in satisfactorily scientific ways. As we shall see, the resolution had to await the emergence of cell biology and the maturation of evolutionary theory.

---

[13] The effects of the vital force are the manifestations of life itself. But these manifestations are the basis for inferring the concept of 'vital force' in the first place. So the reasoning is circular. 'Vital force' provides a mere pseudo-explanation or biological phenomena. It only suffices to label our ignorance.

Chapter 9
Cell Theory and Experimental Physiology: New Ideas in a Changing Society

The metamorphosis of Western Europe during the 18th and 19th centuries illustrates the profound relationships among technology, belief and the wider culture that are typical of human societies (Chapter 2). Between the publications of Lavoisier and Pasteur cited in Chapter 8, there was a massive cultural shift: a long, devastating pan-European war, the industrial revolution with its concomitant demographic, social and economic transformations, colonial expansion, new political movements, and so on. Not surprisingly, the style of thought associated with the Enlightenment and the rise of natural and experimental philosophy was challenged and altered. The character and concerns of science, including biology, changed accordingly.

Hume: the Achilles Heel of Empiricism

David Hume (1711–1776) lived through the Age of Reason. An inheritor of the Lockean tradition, he nevertheless undermined the certainties of classical empiricism. His work remains disconcerting today, though few of his contemporaries grasped its implications.

In *An Enquiry Concerning Human Understanding*, Hume questioned how we can ever obtain reliable knowledge about the world.[1] He said that *deductive* sciences such as mathematics merely work out the logical consequences of our definitions. However, the generalisations of empirical (*inductive*) sciences are no more than 'psychological habits' or 'customs' acquired when patterns of observations are repeated. Empiricists tell us that we only have direct knowledge of sensations ('impressions'), so our belief that there is a world of objects and people ('things in themselves') outside our skins is speculation and may be illusory. The fact that inductive reasoning 'usually seems to work well' does not justify it: induction cannot be justified inductively! Hume pointed out that no possible number of instances

---

[1] This sceptical position was not new in itself; Algazel, William of Ockham and George Berkeley (1685–1753) had all taken comparable standpoints. But Hume developed it in a new and unsettling way.

P.S. Agutter, D.N. Wheatley, *Thinking about Life*,
© Springer Science + Business Media B.V. 2008

of the same observation could lead to certainty. We cannot exclude the possibility of black swans merely because every swan we see is white.

> The contrary of every matter of fact is still possible; because it can never imply a contradiction, and is conceived by the mind with the same facility and distinctness, as if ever so conformable to reality. *That the sun will not rise tomorrow* is no less intelligible a proposition, and implies no more contradiction, than the affirmation, *that it will rise.*[2]

This simple argument had drastic implications. If reliable knowledge can only be obtained by inductive generalisation from observations, but induction can never produce certainty, then (supposedly) reliable knowledge cannot be certain. Our attributions of 'cause and effect' are nothing more than 'custom'. In an age that had come to accept Newton's account of the universe as Eternal Truth and believed, with Locke, that Newton had proceeded as a good Baconian empiricist, Hume had created fundamental difficulties for science – and also, potentially, for the social and political optimism of the Enlightenment. Moreover, Hume's scepticism extended to religion. Since the existence of God can be demonstrated by neither deductive nor inductive argument, it may not be assumed. As la Mettrie had shown (Chapter 8), atheism had become philosophically possible.[3] Hume, a humane and well-liked man, was never reconciled to religious faith.

Kant

Immanuel Kant (1724–1804) confronted the philosophical dilemma that Hume had unleashed on the late 18th century as follows: first, he accepted Newtonian mechanics as the complete and unerring truth[4]; second, he accepted Hume's strictures on the limits of induction; and third, he acknowledged the evident contradiction. How could observation statements stand as exact knowledge?

   Kant agreed that our senses and mind are the instruments by which impressions of the world are received, absorbed and evaluated. However, he proposed that we are *not* passive recipients of those impressions; the universe is not a clockwork mechanism that imposes its order and its laws. Rather, the observer's mind has an innate structure that organises those impressions. Some principles are known not from sensory experience, but innately: e.g. space is infinite and Euclidean, and the basic rules of mathematics and logic are true. This innate structure enables the mind to assemble Newtonian mechanics from sensory information and establish it as mathematical certainty.[5]

---

[2] An Enquiry Concerning Human Understanding, Section 4, Part 1.

[3] Of course, the *non*-existence of God cannot be demonstrated inductively or deductively, so by the same argument, the non-existence of God cannot be assumed either.

[4] In the *Metaphysical Foundations of Natural Science*.

[5] The general notion that our minds are adapted to 'match' an (ultimately unknowable) external reality is echoed in the modern biological idea that our brains have evolved to make effective sense of the world in which we live. If our minds were not able to match external reality, says this argument, we could not have survived as a species.

Essentially, the *actual* structure and constitution of the universe ('things-in-themselves') is irrelevant to us; our cognitive faculties impose law and order upon it, and that is what we know. The world divides into *phenomena* (sensory impressions organised by the mind) and *noumena* (things in themselves). Of the latter we must remain entirely ignorant, since we cannot gain access to the 'thing in itself', or to God or spirit. That was Kant's attempt to *'delineate the reach of knowledge in order to make room for faith'*. Averroës, Anselm and Aquinas might have approved, but Hume (and Locke) would not!

Kant wrote[6]:

> When Galileo let his balls run down an inclined plane to test a gravity he had chosen himself; when Torricelli caused air to support a weight he had chosen beforehand... a light dawned on natural philosophers. They learnt that our reason can only understand what it creates according to its own design, and that we must coax from nature the answers to our questions ... Observations made without prior plan and hypothesis cannot be connected to... what our reason is looking for.

Nowadays, 'observation statements are theory-laden' is a familiar aphorism. We all accept that experiments and their results are pre-loaded with theory. As Karl Popper (1902–1993) put it, scientists evolve myths about Nature and write them up as stories; and when these are submitted to Nature, we expect her to weed out the errors (falsification). Thus, the whole notion of 'inductive science' is false. It is not logically possible to derive a theory from non-committal observations. However, Kant's imagined observer is 'apart from' the universe that he observes; his 'detached mind' is not subject to the laws of mechanics. This persuades us that the observer is 'absolutely at rest', entrenching the Newtonian/Kantian notions of absolute space and time, the metaphysical basis of classical mechanics that Mach challenged in the closing years of the 19th century and Einstein replaced in 1905. (Einstein said of his own *Gedanken* experiments that one could only understand the problem if one could step outside the universe.)

In his *Critique of Judgment*, Kant directly addressed the problem of purpose in biology. He distinguished between 'external adaptation' and 'internal adaptation'. The former is between the organism and its environment, e.g. between a plant and the soil in which it grows. The latter is among the anatomical parts of an organism, or between the organism and its function. Adaptation between organism and function entails Aristotle's 'final cause': organisms act as though they were produced for a purpose. But purposes cannot be *demonstrated*. Therefore, the teleological concept of final cause is in the same category as 'the ideas' (the world, the soul, God); not part of the *phenomena*, but known through the internal structure of the mind.

Kantian and 'Neokantian' philosophy became and remained the dominant perspective on science in continental Europe. Its main impact on biology came through the *mechanistic materialism* of the 1840s and later (see below). It also found adherents in Britain during the first half of the 19th century, largely thanks to the forceful and charismatic William Whewell, founder of the British Association for the

---

[6]From the *Critique of Pure Reason*, translated by Kemp Smith. By all accounts, Kant was an excellent lecturer, clear and lucid; but these qualities did not extend to his writing. The *Critique* is a notoriously obscure and difficult book, so conflicting interpretations of Kant abound.

Advancement of Science.[7] Whewell said that scientific inquiry must emphasise the discovery of 'true causes' by amalgamating disparate phenomena (*'consilience of inductions'*) under a single unifying 'Conception of the Mind', of which the prime example was Newton's universal law of gravitation.

The Romantic Reaction

The career of the poet and painter William Blake (1757–1828) spanned the American and French Revolutions and the early industrial revolution. Blake was an early exponent of Romanticism, a diffuse but intense intellectual reaction against the increasingly dominant 'scientific' worldview, which in his opinion constricted the human spirit. He associated it with three famous English pioneers of natural philosophy: *'Bacon, Newton and Locke, who teach despair to the nations, who teach doubt and experiment'*.

Blake was not alone in his antipathy to natural philosophy. Most Romantics rebelled against the notion that the cold objective evidence of the senses constituted the whole of valid human experience and knowledge. They emphasised the imagination and the emotional rather than intellectual responses to sensation. Many of them held extreme political views: Blake supported the American rebels, the French Revolution and British radicals such as Tom Paine; Wordsworth was active in Paris during the Revolution. Mainly, however, Romanticism was about the conception of art as an expression of individual feeling rather than mere craft, and of poetry as the original, vital human language. It entailed a deep interest in folk-song and the 'voice of the people'. This explicitly opposed the pronouncements of Locke, the Royal Society, and indeed Kant.

Paradoxically, Romanticism influenced some strands of 19th century science, not least biology, mainly in the form of *Naturphilosophie*.

Naturphilosophie

Shortly after Kant's major works were published, a tradition of 'German Idealism' flourished. One of its exponents, Friedrich von Schelling (1775–1854), took issue with Kant's failure to explain how a free, knowing, non-determined subject (the

---

[7] Whewell lived when the word 'science' was taking on its modern signification. It was Whewell who proposed the term 'scientist' for a practitioner of natural philosophy. The term found little favour. When it resurfaced in the later decades of the 19th century, British academics took it for an Americanism and regarded it with contempt; T. H. Huxley epitomised it as 'that abominable trisyllable'. The label was further debated in the 1920s and 1930s but 'scientific worker', 'man of science' etc. remained the preferred usages. 'Scientist' did not become an accepted part of the vocabulary until after the Second World War. Whewell was also responsible for several other additions to our vocabulary, including *physicist, cathode* and *anode*.

observer) can arise from a Nature that is wholly governed by deterministic laws. Schelling tried to resolve this problem by asserting that Nature, including ourselves as observers, constitutes a single and self-forming unity, with an innate organising principle that struggles towards self-consciousness. That single formative energy is the soul or inner aspect of Nature. Schelling's *Naturphilosophie* was attacked and even ridiculed by many of his successors, but it stood against the Kantian monolith. Some biologists saw it as a way of assimilating the Neoplatonist 'ladder of nature' into modern science.[8] It was developed particularly by the remarkable polymath Johann Wolfgang von Goethe (1749–1832).[9]

Goethe

Goethe insisted on both the unity of Nature and the unity between knowledge and feeling. He opposed any specialisation of interests, and similar attitudes informed both his artistic and his scientific output. A pioneer of Romanticism, he wrote some 14 volumes on scientific topics. Science, he asserted, should present to the mind what poetry presented to the imagination and the feelings. His most famous work, *Faust*, expressed his conviction that true understanding depends on the simultaneous acceptance of apparently incompatible points of view rather than a search for compromise. He was one of the first writers seriously to attempt a history of science, in which he emphasised the underlying complementarity of contrasting evidence and beliefs.

Goethe agreed with Kant about the rejection of classical empiricism, but he followed Schelling in dismissing Kant's dichotomy between Nature and Reason and his 'pigeonholing' of human concerns. For example, he could not accept that morality and aesthetics belonged to Reason alone, and not at all to Nature. More significantly, he was critical of Newton and of classical mechanics. Thus, his mature thought stands in self-conscious opposition to Kant.

Goethe was more a contributor of ideas than actual scientific accomplishment, though his writings ranged over botany, comparative anatomy, zoology, physiology, geology, mineralogy, osteology, and his celebrated theory of colours. His best-known scientific discovery, the intermaxillary bone, was made independently by others; but significantly, he was led to that discovery by the notion of harmonious (teleological) evolution of body forms. Thus, he set an alternative trend to Kantian mechanism. His teachings emphasised the distinctiveness of living organisms in respect of principles of organisation and 'purposiveness'. In that sense, he gave Aristotelianism a new lease of life.

---

[8] For example, Christian Gottfried Ehrenberg (1795–1876), probably the first microscopist to witness the division of a cell nucleus, believed explicitly in the 'ladder of nature' and in *Naturphilosophie*.

[9] Goethe was acquainted at first hand with the anti-mechanistic stance of Johann Reil (Chapter 8), since Reil had served as his physician.

Goethe's approach to science emphasised the unification of outwardly dissimilar phenomena. For example, he imagined a discipline that he labelled 'morphology', the systematic study of formation and transformation processes (which would encompass clouds, rocks, plants, animals, colours and human culture) *as they presented themselves to the senses*. Like most of his scientific ideas, this proved sterile; yet it may have contained the germ of great syntheses such as Maxwell's mathematical unification of the studies of electricity, magnetism and optics. Goethe's *Metamorphose der Pflanzen* of 1790 emphasised the similarities within and among plant forms – '*all parts of a plant are modifications of a type-leaf*' – and echoed the belief in the underlying unity of Nature advanced during the first decade of the 19th century by Etienne Geoffroy St Hilaire (1772–1844). Goethe's stance on the disagreement between Geoffroy and George Cuvier (1769–1832) was later recognised and approved by Darwin.

Geoffroy versus Cuvier

Which comes first in biology: organisation or activity? Is function the consequence of form, or is form the manifestation of function? Aristotle had wrestled with this problem and had concluded that form emerges concomitantly with function (purpose) as a result of entelechy. He had also identified structures that were homologous among different animal groups (e.g. the lateral fin of a fish, the wing of a bird, the forelimb of a mammal). Geoffroy, an exponent of the 'formal' approach to biology that was associated with Buffon and later with Goethe, also recognised homologies, though he was inclined to push the concept too far. For Geoffroy, *form determines function*.

Geoffroy invited Cuvier to Paris in 1795, and Cuvier later became professor of anatomy at the National Museum of Natural History. Cuvier's achievements are hard to overestimate. For example, he founded the study of vertebrate palaeontology more or less single-handedly, and he radically revised Linnaean taxonomy. His geological study of the Paris basin with Alexandre Brongniart established the basic principles of biostratigraphy. A champion of natural philosophy and opposed to any hint of Aristotelianism, he was a committed 'functionalist'. For Cuvier, *function determines form*.

The debate between the two men was life-long, though they respected each other. Their disagreement turned on the number of different 'archetypal forms' of organisms: Geoffroy argued for few and Cuvier for many. It is obvious why Goethe was so enthusiastic about Geoffroy and correspondingly antipathetic to Cuvier.

Müller

Johannes Peter Müller (1801–1858) was a powerful, inventive and communicative personality who, deeply influenced by Goethe, was concerned with objects *as they presented themselves to the senses*. His *Handbuch der Physiologie des Menschen*, published between 1833 and 1840, opened a new phase in the study of physiology. It combined comparative and human anatomy, chemistry and physics in the investigation of physiological problems. The most novel part of the *Handbuch* dealt with the action of nerves and the mechanisms of the senses, showing for the first time that the sensation we experience when a nerve is stimulated depends only on the nature of the sense-organ, not on the mode of stimulation. For example, mechanical stimulation of the optic nerve produces the same sensation as a flash of light on the retina.

Like Bichat and some of his French contemporaries, Müller regarded organisation and development as the primary issues in biology. He has sometimes been deemed a 'vitalist', perhaps on account of the following sentence from his *Handbuch*:

> Though there appears to be something in the phenomena of living beings which cannot be explained by ordinary mechanical, physical or chemical laws, much may be so explained, and we may without fear push these explanations as far as we can, so long as we keep to the solid ground of observation and experiment.

However, the most plausible interpretation of the word 'phenomena' in this context is Kantian, and there is no suggestion of radical incompatibility with physics, merely an admission of ignorance. Observation and experiment sometimes yielded data that could not wholly be explained in terms of the chemistry and physics of the early 19th century, but that did not mean that they would *never* be scientifically explicable. Nevertheless, Müller devoted himself increasingly to the study of comparative anatomy as he grew older, in order to understand organisation and development better.

Müller was a very influential teacher. His students included such distinguished scientists as Emil du Bois-Reymond (1818–1896), Hermann von Helmholtz (1821–1894), Carl Ludwig (1816–1895), Theodor Schwann (1810–1882) and Rudolf Virchow (1821–1902). The work of this younger generation coincided with the industrialisation of the German states, and with the incipient movement towards German unification and nationalism. In their hands, biology and medicine were to undergo rapid advances, and the two centuries-old debate between Descartes and Harvey was to be reborn in a new guise.

The Birth of Cell Biology

'Cells' or 'globules' had been seen in biological specimens by many users of the microscope since Hooke. However, not until the dramatic improvement in microscope optics by Lister's achromatic lens (1827) did the idea of the cell as the 'fundamental unit of life' come into being. Hard on the heels of this invention (1827–1828), François-Vincent Raspail (1794–1878) asserted that all animal as well as plant tissues are made up of cells. He also suggested that disease processes are initiated at the cellular level, anticipating Virchow by almost three decades, and he seems to have coined the phrase *omnis cellula e cellula* (all cells come from pre-existing cells)[10] as early as 1825. The great anatomist and surgeon, John Goodsir (1814–1867), accepted Raspail's work and identified the cell as the 'centre of nutrition'.

The period 1835–1840 saw the emergence of 'cell theory' in other European countries, particularly Germany.[11] Müller recognised that cells in plants and animals are essentially similar; various workers observed that the nucleus divides before the rest of the cell; the notion that cells are formed by the coalescence of amorphous material gradually yielded to the belief that cells always arise from pre-existing cells. By the late 1830s, the botanist Matthias Schleiden (1804–1881) had asserted that all cells have a common principle of origin and Theodor Schwann, a student of Müller, agreed. Schwann further asserted that

- All tissues are composed of cells,
- Each cell has an independent ability to live, and
- The total activity of an organism is the sum of the activities of its component cells.

Schwann, like Descartes, blended materialism with Deism. Unlike many of his peers, such as du Bois-Reymond and Helmholtz, he was highly religious[12] and he was influenced, via Müller's teaching, by *Naturphilosophie*. In a passage typical of his split Kantian-versus-*Naturphilosophie* beliefs, he wrote:

> Either the organism is endowed with a force that forms it as a whole according to some idea, or… is subject to forces that act according to blind necessity, forces that are inherent in matter itself.

He declined to make a final judgment on this issue, but he admitted that, if pushed, he would choose to fall on the mechanist side of the fence.

However, the old problem of mechanism remained: if the workings of organisms involve 'nothing but' physics, how can we account for their *purposive* character? The deeply religious Schwann, like the older school of mechanists, transferred purpose from biology to the world as a whole and attributed it to the Creator. Thus, he

---

[10] This phrase may owe something to Harvey's *omnis viva ex ovo* (Chapter 6). The traditional German founders of 'cell theory', Schleiden and Schwann (see following text), were aware of Raspail's pioneering discoveries but held them in contempt: an illustration of the antipathy between French and German scientists that persisted throughout much of the 19th century. Famously, the philosopher Herder had advised German academics to '*Spit the green slime of the Seine from their mouths*'.

[11] As the German states began to move towards unification and the spirit of nationalism grew stronger, German science became animated by the conviction that it was knocking on the doors of Ultimate Truths. The struggle to raise it to world leadership inspired all schools of thought. The often bitter controversies among scientists of the period betrayed an impulse to advance (German) science. Methodical, reductive, mechanistic science became generally perceived as the great beneficiary as well as the great explainer. In the later part of the century, Haeckel expressed the general opinion in his highly publicised view that science would solve *all* mankind's problems during the following two generations. It was the Enlightenment conviction in new nationalistic garb.

[12] Schleiden was such another; in 1863 he published a vitriolic attack on the anti-religious trend of German science.

presumed that organisms were only *quantitatively*, not qualitatively, more purposeful than a mechanical system such as the Solar System. Despite its strong taint of *Naturphilosophie,* this solution to the problem of purpose was broadly adopted by several materialists during the following decade.[13]

Another of Müller's students, Rudolf Virchow, became the founding father of cellular pathology. He acquired the principle, if not the wording, of *omnis cellula e cellula* from the writings of John Goodsir, and hence (without acknowledgment) from Raspail. He dedicated his *Cellular Pathology* of 1858 to '*John Goodsir FRS, Professor of Anatomy in Edinburgh, the earliest and most acute observer of cell-life*'. Bichat had insisted that the roots of organisation and purpose lay in tissues rather than organs (Chapter 8). Virchow extended this analysis to another level: he traced those roots, and the roots of malfunction (disease), to the cells, the 'atomic units' of biological organisation. Virchow's authority was instrumental in making the cell the centre of attention among biologists. He regarded the cell as the physiological as well as the morphological unit of life and emphasised its continuity in development.

Mechanistic Materialism

There was a sharp philosophical division between Virchow and other students of Müller – du Bois-Reymond, Ludwig and their colleagues – who took a strongly Kantian stance, repudiating *Naturphilosophie* and the views of their teacher.[14]

Mechanistic materialism was articulated by the Berlin Physical Society, founded in 1845 and originally dedicated to explaining all aspects of life (including mental processes) in terms of Newtonian mechanics. Du Bois-Reymond and his colleagues were particularly concerned to apply this approach to physiology. At the time, Dalton's atomic theory was only 40 years old and modern understanding of chemical bonds did not exist. Later, when the original project of mechanistic materialism failed (because the physics of the day could not explain all of physiology), chemistry had matured sufficiently to redirect that project. Biochemistry was to become the phoenix that rose from the ashes of the Berlin Physical Society's endeavours.

---

[13] In 1842, the mechanist Lotze attacked the assumption that living matter is specifically distinctive and upbraided the vitalists for their obscure use of the term 'force', but he too had recourse to Divine intervention to explain 'purpose'. Even Vogt, the most extreme of materialists, followed that path.

[14] Du Bois-Reymond invited Virchow to join the Berlin Physical Society; Virchow declined. The two men appear to have respected each other, but their philosophical and scientific differences could not be reconciled. Both were political activists; Virchow in particular was a liberal social reformer and an implacable opponent of Bismarck, who once challenged him to a duel. (The challenge was declined.) Interestingly, however, du Bois-Reymond's closest colleague, Brücke, perceived the cell as an 'elementary organism', highly complex in structure. This Virchow-like stance helped to inspire subsequent studies of intracellular structure.

The extreme mechanistic materialist position had been outlined as early as 1841, when du Bois-Reymond quoted the French materialist physiologist Henri Dutrochet[15] (1776–1847) approvingly:

> The more one advances in the knowledge of physiology, the more reasons one will have for ceasing to believe that the phenomena of life are essentially different from physical phenomena.

A year later, du Bois-Reymond and Ernst Brücke (1819–1892) had sworn to

> validate the basic truth that in an organism, no other forces have any effect than the common physicochemical ones.

Elsewhere, we find:

> a vital phenomenon can only be regarded as explained if it has been proven that it appears as the result of the material components of living organisms interacting according to the laws that those same components follow in their interactions outside of living systems.

Helmholtz had been recruited to this cause by the end of 1845, and Ludwig by the early part of 1847.

The preface of du Bois-Reymond's *Researches on Animal Electricity* (1848–84) is virtually a mechanistic materialist manifesto,[16] expressing the views of all four men, predicting the necessary and complete dissolution of physiology into physics. The experimental physiologist must proceed with physical and mathematical exactness, always seeking to explain biological phenomena in terms of mechanics. Du Bois-Reymond defined 'mechanics' as the motion of particles of matter:

> if only our methods sufficed, an analytical mechanics of the general life process would be possible... All changes in the material world within our conception reduce to motions. Therefore even that process cannot be anything but motions...

Thus, the mechanistic assertions of 1847 echoed those of Boerhaave and la Mettrie a century earlier. Du Bois-Reymond and his colleagues were inflexibly anti-vitalist, insisting on non-mystical causality for all living processes; they fostered the use of observation and experiment; and they insisted that the attempt to reduce physiology to physics was practicable and potentially useful.

There is no doubt that du Bois-Reymond, Brücke, Helmholtz and Ludwig made invaluable contributions to physiology. Their work drew worldwide public attention to their 'cause' and disseminated the spirit of anti-vitalism and experimentalism in physiology. They used physical methods and concepts and mathematical techniques to the limit of their powers, and to considerable effect. But their real contributions, especially those of Ludwig, the doyen of experimental physiologists, lay in specific factual discoveries in aspects of anatomy and histology and, later, chemistry.

---

[15] Dutrochet discovered cells in plants, described respiration, light-sensitivity and embryo development in plants, and also discovered osmosis. His approach to science was 'mechanistic', unlike that of Bichat or many of his contemporaries in France.

[16] His philosophical views were also set out in a series of essays: The Limits of Natural Science (1872).

Once again, however, they could offer no satisfactory answer to the problem of purpose. For du Bois-Reymond and his followers, Schwann's (theistic) solution to that problem was unacceptable. Rather than offer any alternative they dismissed and even ridiculed the issue of 'purpose'. Since in du Bois-Reymond's opinion the mechanistic doctrine *had* to prevail, organisation, development and purposiveness were simply disregarded as matters unworthy of serious scientific discussion.

Helmholtz's scientific works combined philosophical insight, exact physiological investigation, mathematical precision and a deep grasp of physical principles. After his early collaboration with du Bois-Reymond, he adopted a position closer to British positivism (see Chapter 12) than to mechanistic materialism. That is to say, he became philosophically opposed to Kant's conclusion that time, space and causation were mental structures through which the world is comprehended, and returned to the empiricist view that all knowledge comes from the senses.

An Alternative Tradition of Physiology

Before the end of the 19th century, mechanistic materialism in its original form had faltered. Ludwig and his student Adolf Fick agreed that physiology was *not* wholly reducible to mechanics after all. Nevertheless the spirit of mechanistic materialism survived, inspiring the early genetic work of Thomas Hunt Morgan (1866–1945), an influential book by Fick's student Jacques Loeb (1859–1924),[17] and the pioneers of biochemistry. The mechanistic materialists had spread their influence throughout Europe and North America thanks to the numerous students they (especially Ludwig) had trained. Many of those students became professors of physiology.

There was a parallel tradition of experimental physiology in 19th century France that was entirely divorced from mechanistic materialism. François Magendie (1783–1855), a contemporary of Dutrochet, was considered the founder of this tradition. His most famous student, Claude Bernard (1813–1878), was dubbed 'physiology itself' by Pasteur. Bernard accomplished a remarkable volume of research on the pancreas, the liver, smooth muscle tissue and brain lesions, leading to crucial insights into metabolism and its control. He had no philosophical commitment or programme comparable to that of du Bois-Reymond and his allies, and his work shows the influence of Bichat's pioneering studies of histology as well as the vivisection skills of Magendie and the scientific methods of Dutrochet. Towards the end of his career he became increasingly impressed by what we now call *homeostasis*, the set of mechanisms by which body water content, body temperature, blood pressure, blood glucose concentration and a host of other variables are maintained within narrow limits.[18] This focus was quite different from any that derived from mechanistic materialism.

[17] The Mechanistic Conception of Life. University of Chicago Press, Chicago, IL (1912).

[18] The word 'homeostasis' was introduced by the American physiologist Walter Bradford Cannon (1871–1945) in the late 1920s.

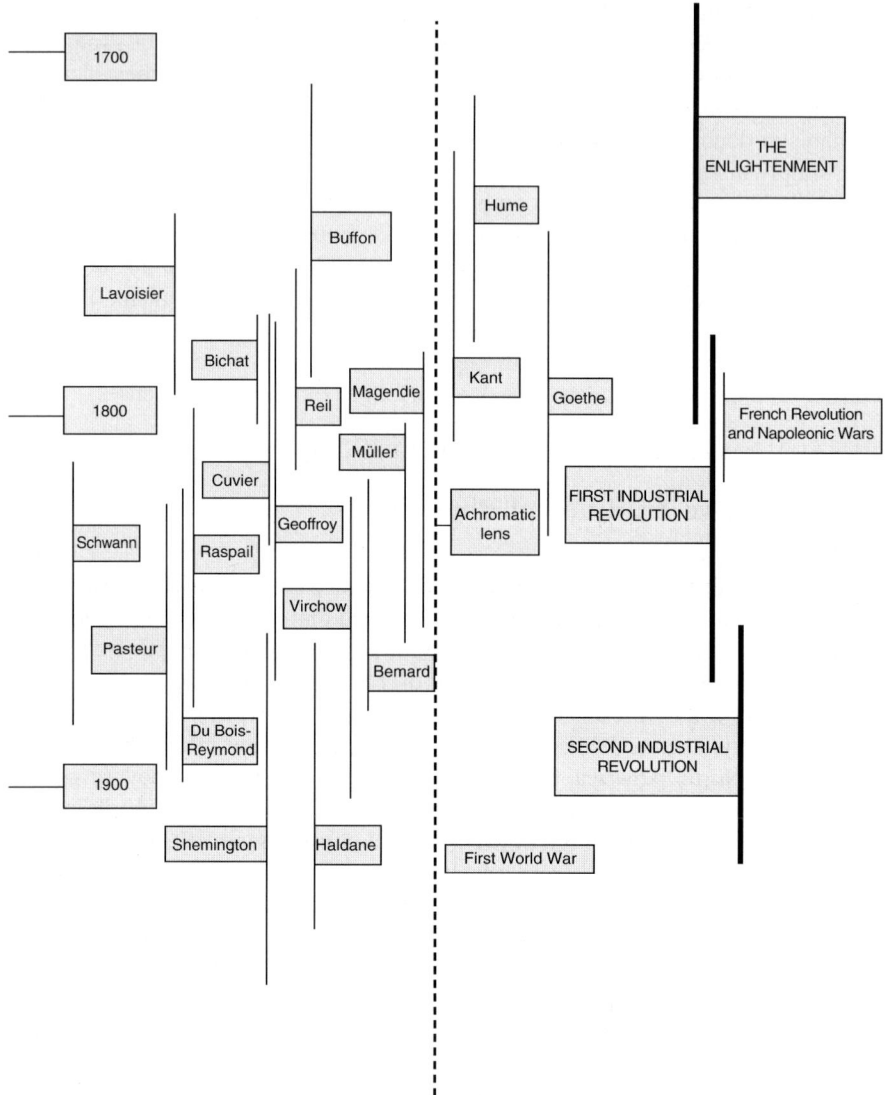

Fig. 9.1 Time-course of some salient contributions to thought and of events related to the development of biology during the period 1700–1950, highlighting the dates of the Enlightenment and the First and Second Industrial Revolutions. This diagram may be seen as a continuation of Fig. 8.1

There was a clear gulf between the tradition of French experimental physiology and the mechanistic materialism of the post-1847 Berlin school. The so-called 'holistic materialism' or 'organicism' of early 20th century physiology can be seen as a resolution of this conflict. But holistic materialism was not a single body of belief. The position taken by John Scott Haldane (1860–1936) was close to that of Virchow, while contemporaneous physiologists such as Charles Sherrington

(1857–1952) and Walter Cannon had subtly different views. In opposition to the mechanistic materialists, they held that the whole was greater than the sum of its parts, i.e. that emergent properties in biology are real and essential for understanding. However, they still believed that the study of the parts *rather than* of the whole would ensure the advancement of knowledge, while Haldane favoured the study of the parts *as well as* the whole.

Later in the 20th century, the 'parts' became cells and fractions of cells rather than organs and tissues; biochemistry colonised the territory of experimental physiology. The emphasis on the study of parts was redirected. Cell theory and experimental physiology, which at their inception in the 19th century had seemed distinct, became amalgamated (Fig. 9.1).

Implications for the Development of Biology

By the middle of the 19th century, science had been elevated to a position of cultural pre-eminence by the huge social changes attendant on industrialisation and colonial expansion. The foundations of chemistry, and technical innovations such as the achromatic microscope lens, had fostered a rapid growth of knowledge about organisms. Out of the concurrent debates between mechanists and anti-mechanists, Kantians and proponents of *Naturphilosophie*, a new scientific biology was emerging.

The pioneers of cell theory pronounced that all organisms consisted of one or more cells, each a 'centre of nutrition' and of replication, and that all cells arose from previously existing cells. So cell theory represented, and fostered, a belief in the *unity* and the *continuity* of life. It changed the direction of biology. Yet as we have seen, it did not resolve the debate between mechanists and anti-mechanists, nor did it solve the 'problem of purpose'. Those achievements would require a mature theory of evolution. Evolutionary theory, like cell theory, fosters belief in the *continuity* of life, but its fundamental concern is a rational explanation of life's *diversity* rather than its unity.

The origin and maturation of the theory of evolution is a complex topic, but it was crucial for the emergence of a fully scientific biology. To understand it fully, we first need to consider two other topics that were debated from the late 17th century until the late 19th: embryo development and spontaneous generation. Both these topics induced two centuries of heated speculation and innovative experiments, and cell theory played a significant part in finally resolving the debates. The history is fascinating in itself and it throws further light on the emergence of modern biology.

Chapter 10
Embryos and Entelechy

Among the many features of organisms that inspire our sense of wonder, the emergence of new life stands at the pinnacle. No one can witness the birth of a baby without a feeling of awe, and that feeling extends to non-human species. The germination of seeds and the growth of plants are commonplace events but they are still marvels. Every spring, a hawthorn hedge bears thousands of identical flowers and tens of thousands of new leaves, each a miniature jewel of complex form. Similar examples abound in the natural world. Where do these new manifestations of life come from, and what brings them into being in such profusion and with such seemingly mechanical regularity? 'Miracle' is the wrong word for these occurrences; 'miracles' are one-off events that flout laws and therefore lie beyond the purview of science (unless and until a rational explanation is found), while the development and maturation of embryos are aspects of the uniformity of nature and follow law-like, regular patterns. But how are they to be explained scientifically?

Aristotle proposed an *epigenetic* account (Chapter 7), which was inherently teleological and could not be reconciled with the spirit of the Scientific Revolution. In consequence, many mechanists after Descartes took a radically different view of embryo development, *preformationism*, which appeared to be wholly non-teleological. Meanwhile, anti-mechanists, following Harvey, continued to consider epigenesis more plausible. The ensuing debate between these two traditions was long and complex, raising deep philosophical issues with implications for biology as a whole. This aspect of history is central to the 'war of attrition' that freed biology of Aristotle's influence.

Historically, the idea of 'purpose' is more deeply entrenched in embryology than in any other branch of biology. Aristotle's *De Generatione Animalium* implicated entelechy first and foremost in the formation and maturation of embryos, 'actualising the potential' of unformed reproductive matter (Chapter 7). The central challenge during the past 350 years has been to account for this inescapably purposive process in non-teleological terms. Recent advances in knowledge and techniques have enabled us to do so, though major questions remain to be answered.

P.S. Agutter, D.N. Wheatley, *Thinking about Life*,
© Springer Science + Business Media B.V. 2008

Preformationism

According to this view, the new individual pre-exists in miniature within what we would now call the gametes of one or other parent. Fertilisation *occasions* the development of a new individual but does not *cause* it. Development consists only of growth from the miniature to the mature size; there is no 'coming into being out of non-existence', no emergence of novel anatomical structures from 'unformed reproductive matter'. By thus evading any discussion of 'cause', preformationists dodged the problem of entelechy.

For mechanists in the late 17th and early 18th centuries, therefore, preformationism was the hypothesis of choice. Swammerdam made the first explicit statement of it in the late 1660s; this was expounded in more detail by the Cartesian Nicholas Malebranche (1638–1715) in 1674.[1] Nicolaas Hartsoeker (1656–1725) published a drawing of a 'homunculus' cramped inside a spermatozoon, an imaginative construction that has been reproduced in many modern biology textbooks (Fig. 10.1). Preformationism had clear advantages:

- It 'explained' how the parts of organisms came to be integrated in structure and function. For example, since it was believed that the heart could not beat without nerves and the nerves could not exist without the heart, it seemed necessary to assume that heart and nerves appeared simultaneously rather than sequentially in the embryo. How could epigenesis explain such simultaneity?
- It was compatible with some developments in 17th century theology, particularly Calvinism and Jansenism, traditions that followed Augustine in supposing an omnipotent God and a passive Nature.
- It could be (and was) reconciled with the work of the early microscopists (Chapter 6), some of whom believed that miniature adult organisms could be seen in the reproductive organs of adults.

However, there were at least three variants of the preformation idea. Some of its exponents held that the pre-existing miniature forms were encased in the ovaries of the female. Others held that they were encased in the spermatozoa in the testes of the male (recall that spermatozoa were discovered by Leeuwenhoek in 1677). Still others assumed that preformed 'germs' had existed in the soil since the Creation. Organisms were supposed to take in these germs with food. In the *appropriate* organism under the right conditions, the germs entered the ovaries, were fertilised, and developed into new individuals. This variant of preformationism was first proposed by Claude Perrault (1608–1680)[2] but attracted few followers until the 1760s.

---

[1] *Recherche de la vérité*. Published in English as Lennon TM, Olscamp PJ (eds.) (1997) The Search After Truth. Cambridge University Press, Cambridge.

[2] It is strikingly similar to the idea proposed by St Augustine and sustained by mediaeval scholastic writers who belonged to the Plato-Augustine tradition. It was expounded by the physician James Cooke (1762) A new theory of generation, according to the best and latest discoveries in anatomy; Buckland, Dilly, Keith and Johnson. London. Cooke's account has an archaic tone, in contrast to the 'modern' styles of Boerhaave, la Mettrie and indeed Buffon.

Fig. 10.1 A 'homunculus' inside a spermatozoon. The imaginative drawing by Nicolaas Hartsoeker in his *Essai de Dioptrique*; Paris, Anisson, 1694

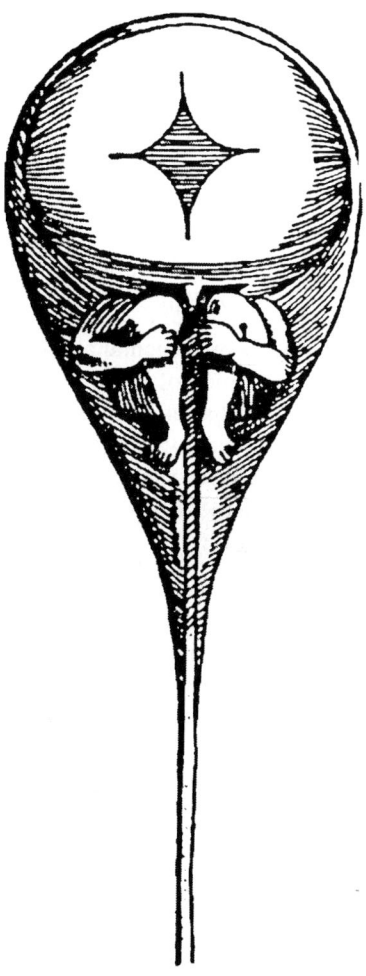

Problems with Preformationism

Although it evaded the teleology of Aristotelian epigenesis, preformation gave rise to difficulties of its own. One important problem was that if every individual was preformed in a parent, no species could have had an origin in historical time. The solution was to attribute the origins of species to divine action when the world was first created. For this reason, preformationism was fundamentally incompatible with any form of evolutionary theory, so its pre-eminence during the period *c*.1670–*c*.1750 partly explains why 'transformism' (evolution) was not seriously considered until later in the 18th century.[3] Proponents of preformation, like other mechanists of the period, were obliged to refer all intractable problems in biology and medicine to God. That 'solution' was incompatible with the spirit of the Scientific Revolution and was rejected by atheists such as la Mettrie.

Of course, those who believed in spontaneous generation (Chapter 11) could not accept preformation, the two beliefs being mutually exclusive.

Preformationism also made it difficult to find mechanistic explanations for a number of common observations, which again were ascribed (perforce) to divine action. For example:

- Gross developmental abnormalities
- Regeneration of lost parts, common in some species
- The fact that offspring resemble *both* parents, not just one
- Geographical variation and racial differences
- Hybrid forms such as mules, which obviously cannot be preformed in the gametes of either horses or donkeys

However, the most telling argument against preformationism was a logical one. Within the gametes of the miniature new individual there must be an even smaller new individual, in the gametes of which there must be a still smaller one. Unless this series continued to infinity, the species would run out of stored 'homunculi' and become extinct. That problem was insoluble. It led ultimately to the demise of preformationism by *reductio ad absurdum*.

Taxonomy and the Critique of Preformationism

All variants of the preformation hypothesis made the concept of 'species' much more rigid than it had been in Aristotle's work or the writings of the Scholastics. This new rigidity facilitated advances in taxonomy. Such advances became necessary as colonial expansion brought knowledge of thousands of new species to Western Europe.

In 1744, Carl von Linné (Linnaeus) (1707–1778) presented a novel taxonomic theory[4]: the present world and its inhabitants arose by descent from a few original forms that had been created by divine action on an equatorial island ('Eden'). Aware of evidence for the apparently sudden creation of new species in the latter-day

---

[3] The 'germ' variant allowed for the appearance of life in historical time, as fossil forms seemed to imply, but it did not entitle its proponents to believe in evolution. The modern reader of these historical texts may be confused by the fact that preformationists were often labelled 'evolutionists' (because the mature individual 'evolved' from a miniature precursor). The meaning of the word 'evolution' in the 18th century was therefore entirely different from the meaning it acquired *c.* 1850 and retains today.

[4] *Oratio de telluris habitabilis incremento* (Oration on the Increase of the Habitated World). Linnaeus's *Systema Naturae*, giving his complete taxonomy of animals, was published 14 years later.

world, Linnaeus suggested hybridisation of the original forms. Such 'origin of species by hybridisation' is not an *evolutionary* concept of species change; indeed, the followers of Linnaeus opposed Lamarck and his successors. It did, however, promote interest in hybridisation and its limits, which was to lead a century later to the classic studies by Mendel (Chapter 13).

Linnaeus's system of taxonomy resembled Aristotle's in that it relied on just a few major characteristics of organisms, e.g. being warm or cold blooded, reproducing oviparously or viviparously. Such 'artificial' systems are quick and economical to use but can lead to serious errors in classification. 'Natural' systems, preferred by Buffon and others, involve many more characteristics, including details of embryo development.[5] They are more cumbersome to use but overall they are biologically more informative. Buffon was the first to conceive of species in terms of *fertile interbreeding*.

Epigenesis Reborn

Whatever the defects of preformationism, a return to epigenesis obviously entailed a commitment to Aristotelian entelechy or to some brand of 'vitalism', i.e. a betrayal of the mechanistic philosophy of the Scientific Revolution. Caspar Friedrich Wolff (1733–1794) nevertheless adopted an Aristotelian position. Considered a founder of descriptive embryology, Wolff said that to explain the emergence of organisms from embryos it was necessary to presume the action of a '*vis essentialis*', an organizing, formative 'force' equivalent to entelechy. Wolff probably did not consider the *vis essentialis* a 'force' in the strict Newtonian sense, but used 'force' as a metaphor for the 'intrinsic tendency' of embryos to mature. In other words, he was not an explicit 'vitalist' like Casimir Medicus. In his debates with preformationists, Wolff relied on observational data, pointing out that early embryos show no adult anatomical structures. The preformationists replied that failure to see anything does not mean that nothing is there. That was a fair answer in an age of inadequate microscope optics. However, Wolff also showed that the earliest visible structures did not resemble mature parts; for instance, the early gut is a flat structure, not a tube. The preformationists had no satisfactory reply.

From the 1740s, efforts were made to establish a 'mechanistic' epigenetic embryology. These endeavours were pioneered by Pierre de Maupertuis (1698–1759). Maupertuis, like Descartes, returned to an idea that dated back to Galen and even to Hippocrates: two 'seeds' must fuse to initiate embryo development.[6] Development could continue only if the particles constituting these seeds attracted

---

[5] Through the use of 'natural' systems, the complex hierarchy of classes of organisms became apparent: living things fall into natural groups. It is difficult to account for this except by a theory of evolution.

[6] Hoffheimer M (1982) Maupertuis and the eighteenth-century critique of preexistence. *J Hist Biol* 15:119–144.

each other. An inertial principle inherent in the particles caused them to arrange themselves into the increasingly complicated structure of the embryo. This 'mechanistic epigenesis' entailed a 'vital' conception of matter, as in Buffon's distinction between the organic and the inorganic (Chapter 8). In other words, it did not *eliminate* the concept of entelechy, but concealed it among the allegedly distinctive properties of organic matter.

In contrast to preformationism, Maupertuis's account allowed the embryo to come into being in historical time. It also entailed *material inheritance*, the transmission of an atomised 'hereditary substance' from one generation to the next. The identity of the species could only be conserved if this substance was *unaltered* during transmission. This implication of 'mechanistic epigenesis' opened the door to species transformation: if the transmitted substance *was* changed, the species would be changed as a result. Buffon developed this idea in his *Histoire Naturelle*. His account of embryo development implied that species were not immutable, but were subject to some degree of modification by environmental influences – mainly 'degenerative' changes. Significantly, Lamarck was a disciple and follower of Buffon (see Chapter 12).

The debate between epigenesis and preformationism did not end with Wolff, Maupertuis and Buffon. For example, it formed part of the disagreement between Cuvier and Geoffroy St Hilaire (Chapter 9). Cuvier's antipathy to Aristotelianism and to vitalism made him a preformationist. Geoffroy, seeing the relevance of embryo development to the relationships among groups of species, believed in epigenesis.[7]

The Mammalian Ovum and the Growth of Descriptive Embryology

The improvement in microscope optics in 1827 and the subsequent rise of cell theory transformed the debate about embryo development. Cell theory was instrumental in making embryology into an experimental as well as a descriptive science. The seed-crystal for this change was von Baer's discovery of the mammalian ovum, predicted by Harvey a century and a half earlier. Subsequent microscopic studies killed preformationism in its classical form.

There were precedents for von Baer's discovery, but none had comparable generality or impact. Von Baer inferred from his observations that *all* sexually reproducing animals arise from an ovum and that fertilisation entails the penetration of the ovum by a spermatozoon. Moreover, embryo development follows the same pattern in all vertebrates; the visible structure that forms first is the precursor of the spine. His subsequent studies suggested that in very early embryos, before anything

---

[7] Geoffroy's studies of embryology and particularly of abnormal development led him to consider that species change was not slow and gradual, as Lamarck proposed, but occurred in bursts caused by changes in the ways that embryos developed. This view presaged the 'mutationism' hypothesis of de Vries (Chapter 13) and, in a different sense, the 'punctuated equilibrium' model of evolution proposed in the 1970s by Eldredge and Gould.

resembling an adult structure can be seen, there are three 'germ layers' from which different tissues will later form.[8] Each step in development, observed von Baer, depends on the preceding step(s), but the direction of the whole process is determined by the 'goal state', the finished organism, just as Aristotle had said. Thus, the dawn of cell biology saw a confirmation of epigenesis. The achievement was notable, but it still left the problem of entelechy unsolved. How could a genuinely *scientific* (mechanistic) account of embryo development be reconciled with a process that seemed inescapably *purposive*?

Von Baer's comparative studies led to a number of other important conclusions. He showed that within any major taxonomic group such as vertebrates, early embryos resemble each other closely; only at later stages in development do they diverge. Importantly, he found that these resemblances are among *embryos*, not between the embryos of more complex animals and adult forms of simpler ones; a point that some later writers were to misconstrue. In the generations following von Baer, many studies of embryo development were conducted on a wide range of animal species. Technical improvements in microscope optics, specimen preparation and staining enabled the stages of development to be described in ever greater detail. The most interesting general conclusion was that early embryo stages are very much alike in all animals more elaborate than hydras and jellyfish: the fertilised egg divides several times,[9] the cells become arranged into a sphere called a blastula and then migrate to form the three germ layers, and the embryo takes on an overall structure, the *gastrula,* within which adult tissues and organs begin to form. Only at later stages of development do embryos of different species become anatomically divergent. Thus, von Baer's findings were confirmed, but much detail was added. In particular, *homology* was given a new significance: structures that are homologous in adult forms, such as the fin of a porpoise and the wing of a bat (to choose one of Darwin's examples), are *identical* in the developing embryo.

Darwin's theory of evolution motivated at least some later 19th century work in descriptive embryology: embryos of modern organisms may *recapitulate* developmental stages of their ancestors, which can therefore be 'reconstructed' even in the absence of informative fossils. The best-known example of recapitulation concerns the bones of the mammalian inner ear, which developed from the jawbones of ancestral reptiles (Fig. 10.2). This evolutionary connection was inferred from

---

[8] This concept was to be elaborated almost a century later by Walther Vogt (1888–1941) in a painstaking series of studies using vital dyes.

[9] Cell division was first described by Hugo von Mohl (1805–1872) in 1835. In the same year, Gabriel Valentin (1810–1883) proposed that the nucleus of the cell divides first. Valentin recognised that all animal as well as plant cells contain nuclei. Ehrenberg described nuclear division clearly in 1838 (see Chapter 9). In two landmark papers of 1841 and 1842, Carl Bergmann (1811–1865) appears to have been the first to understand the nature of cleavage in early embryos *and* to recognise that the blastula is made up of cells.

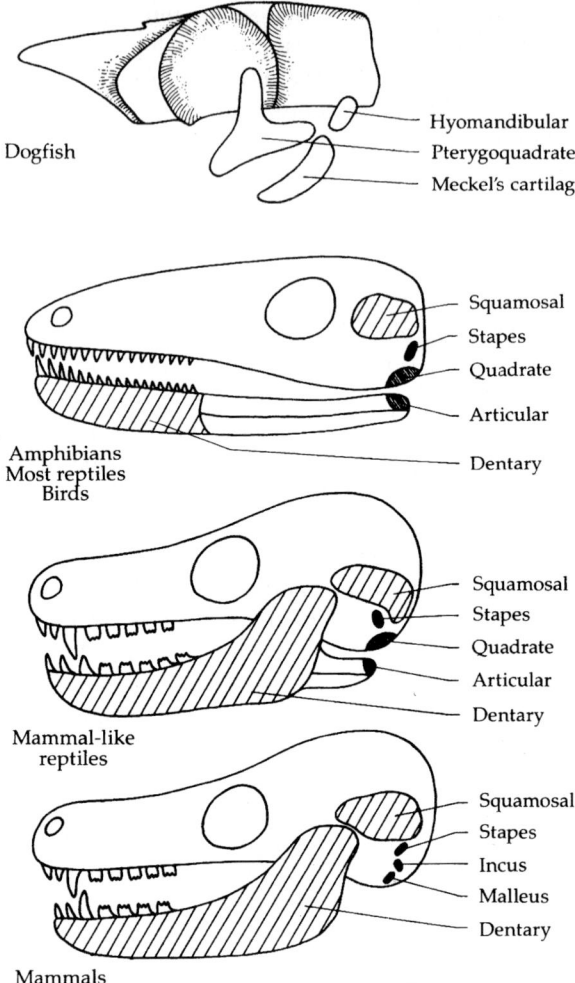

Fig. 10.2 The evolution of the bones of the mammalian inner ear. This sketch illustrates the ancestry of the three inner ear bones in mammals (*malleus, incus and stapes*). In amphibians, most reptiles and birds, the lower jaw is 'hinged' to the skull by two small bones, the quadrate and the articular. In mammals, the enlarged dentary of the lower jaw is 'hinged' to a different part of the skull, the squamosal; the articular and quadrate have now become the malleus and the incus, respectively. The stapes is present in all terrestrial vertebrates; it corresponds to the hyomandibular in the skull of bony fishes. Reprinted by permission of the publisher from Moore JA (1993) Science as a Way of Knowing: the Foundation of Modern Biology, p. 177. Harvard University Press, Cambridge MA. Copyright 1993 by the President and Follows of Harvard College

descriptive embryology before any fossils of mammal-like reptiles were discovered; the fossils confirmed the inference. Thus, the epigenetic account of development not only proved compatible with evolutionary theory, it also became an important source of supporting evidence.

Haeckel and the Recapitulation Hypothesis

Even before Darwin's *Origin of Species* was published, the idea of recapitulation was well established: the taxonomic relationship between two species was echoed in their earliest appearances in the fossil record *and* in their patterns of embryo development. The great Swiss naturalist Louis Agassiz (1807–1873) recognised these correspondences; but in 1857, only two years before the first publication of the *Origin of Species*, he attributed them to the Will of the Creator.[10]

Ernst Haeckel (1834–1919) was a student of Müller and Virchow. When the *Origin of Species* was translated into German in 1860 he was immediately persuaded by it and became the pre-eminent spokesman of Darwinism in Germany. He was professor of zoology at the University of Jena and inspired many studies of embryology and natural history over the next generation, developing Darwinism as a popular movement with social and political overtones as well as a scientific research program. His particular concern was to unify descriptive embryology with evolutionary theory.

Aware of examples of 'recapitulation' from his own research and from the studies of countless others, Haeckel elevated the phenomenon into a 'fundamental law of organic evolution': ontogeny is a recapitulation of phylogeny. As he expressed it[11]:

> The series of forms through which the individual organism passes during its development from the ovum to the complete bodily structure is a brief, condensed repetition of the long series of forms which the animal ancestors of the said organism, or the ancestral forms of the species, have passed through from the earliest period of organic life down to the present day.

Haeckel overstated his case, much as Geoffroy St Hilaire had overstated the case for homology of body forms, speculating far beyond the limits of evidence and inviting rejection of the general thesis. And unlike von Baer, Haeckel seems to have supposed that developing embryos pass through stages resembling the *adult* forms of certain ancestors.

The unification of embryology with evolution and comparative anatomy has long been accepted as a general principle, but many details of Haeckel's claims have been rejected. This has diminished his reputation; but it is worth remembering that he exerted a great and mainly benevolent influence on embryological research in the later 19th century and was himself an indefatigable investigator and writer. He was also the first to propose that the nucleus transmits hereditary information from the parent cell to the daughter cells.

---

[10] An Essay on Classification. Longmans, London.
[11] The Evolution of Man. Eckler, New York, 1905.

Cell Division and the Beginnings of Experimental Embryology

By the 1850s, cell theory was established and a growing consensus accepted the principle *omnis cellula e cellula*. This was the context in which George Newport (1802–1854) made the first explicit observation of fertilisation, the entry of an ovum by a spermatozoon.[12] Newport placed frog eggs in glass chambers just wide enough to accommodate them, then fertilised them with semen on the point of a pin. The fertilised egg divided along a plane through the point of fertilisation; this point later defined the position of the head of the developing tadpole. These experiments related the site of entry of the spermatozoon to the direction of cleavage and to the axis of the developing embryo.

Newport's work can fairly be considered the beginning of experimental embryology. However, largely because of Haeckel's influence, the evolutionary overtones of *descriptive* embryology occupied centre stage for the following 20 years. The next major advances in experimental embryology took place immediately after Ernst Abbe joined the firm of Carl Zeiss and initiated a series of important innovations in microscope technology, beginning with the introduction of the substage condenser in 1873 and continuing with the oil immersion lens in 1878 and the apochromatic lens, which eliminated spherical as well as chromatic aberration, in 1882. The effect on cell biology[13] was almost immediate. Intracellular structures other than the nucleus were identified and the mechanisms of cell division were elucidated. Almost 200 papers about cell division by 80 different authors[14] were published between 1874 and 1878. In particular, *chromosomes* were recognised and became a topic of intensive investigation.

These discoveries were directly relevant to the understanding of embryogenesis and, as will be seen in Chapter 13, of heredity. During the 1880s, Walther Flemming (1843–1905), Edouard van Beneden (1846–1910), Eduard Strasburger (1844–1912) and Theodor Boveri (1862–1915) were prominent in this field. They elucidated the essential facts of cell division and stressed the equal distribution of chromosomes to both daughter cells. In 1882, Flemming described the longitudinal splitting of

---

[12] Martin Barry (1802–1855) had observed spermatozoa *within* ova in 1840, but had not witnessed their entry. In 1845, Rudolf von Kölliker had shown that spermatozoa and ova are cellular products of organisms, and that the new individual forms from the fertilised ovum by cell division. Karl Reichert (1811–1883) observed cell division in the testes in 1847. Thus, Newport's experiments were at the forefront of research. Not until 1875 was it established that fertilization in both animals and plants entails the physical union of the sperm and egg *nuclei*. This was achieved by Wilhelm Hertwig (1849–1922), whose main subject of study was the sea urchin, subsequently a favourite animal for experimental embryologists.

[13] This term was coined by Jean-Baptiste Carnoy (1836–1899) in 1884.

[14] The earliest of this series was an 1873 publication by Friedrich Schneider (1831–1890). Schneider gave a detailed and accurate description of *mitosis*, which we now recognise as the normal method of cell division in multicellular organisms, and showed that it occurred throughout embryo development. But his paper seems not to have been widely read and his pioneering status has seldom been acknowledged.

chromosomes to produce two identical copies. A year later, van Beneden declared that although chromosomes become invisible between cell divisions, they nevertheless persist and are the basis of the continuity between parent and daughter cells. He also showed that the chromosome number is reduced during the special division process, *meiosis*, that produces gametes (spermatozoa and ova). In 1886, August Weismann (1834–1914) predicted that gametes must contain only one of each chromosome pair from the parent cell (a condition described as *haploid*); the full chromosome complement (*diploid*) is restored when the nuclei fuse at fertilisation. This hypothesis was confirmed by the microscopic studies of Boveri in 1888 (Figs. 10.3 and 10.4).

On a more general level, these findings implied that the contents of cells were not randomly arranged but highly organised, as predicted by Brücke. In 1874, William His (1831–1904) advanced the hypothesis that each part of the fertilised egg (*zygote*) corresponds to, and leads to the formation of, a specific part of the mature organism. His's hypothesis would have made no sense if cells had been conceived as disorganised bags of solution, but in the light of advances in cell biology, it was plausible enough to stimulate experimental investigation.

Preformation versus Epigenesis in a New Guise: Roux and Driesch

In effect, what His suggested was a new version of preformationism: although the adult form did not actually *exist* in the zygote (or in the gametes), it was exactly represented by the spatial organisation of material within that cell. In the hands of Wilhelm Roux (1850–1924), this notion led to the hypothesis of *mosaic development*. Mosaic development differed in a fundamental way from earlier versions of preformationism: it was *not* fundamentally incompatible with evolutionary theory, which by the 1870s was well established in Germany. In any case, Roux was a student of Haeckel and deeply committed to Darwinism. Nevertheless, it was a new attempt to explain embryo development mechanistically, avoiding any hint of entelechy. Indeed, Roux called his approach 'developmental mechanics'.

Roux placed embryology on a sound experimental footing, though some of his inferences were premature and his results had subsequently to be reinterpreted. His earliest studies, in which he continuously rotated fertilised frog eggs to eliminate the directional effects of gravity and other external forces, proved that animal embryos are self-determining; their development is not affected by the environment. (That may seem obvious, but Roux was aware of the effects on plant growth of soil chemistry, light, gravity, heat and other factors, and it was reasonable to ask whether animal development might be subject to similar influences.) He also confirmed Newport's findings about the plane of first cleavage and the main axis of the embryo. He then went on to test His's hypothesis by waiting until the fertilized egg had divided into two or four cells and then destroying one of those cells with a hot needle. As His would have predicted, half-embryos resulted in the early stages of development. But after the gastrula stage, the embryos 'completed' themselves

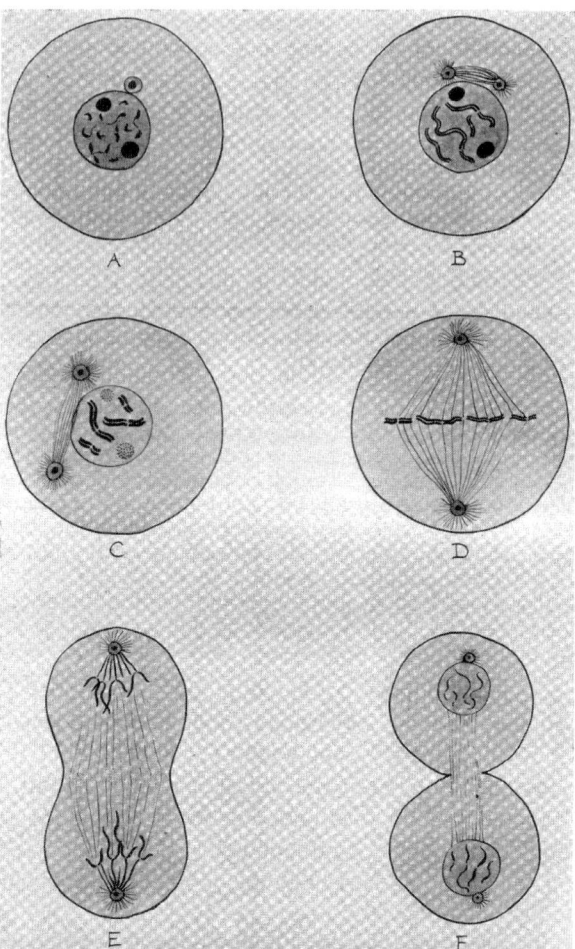

Fig. 10.3 Division of an animal cell: mitosis. (A) represents an non-dividing (*interphase*) cell. The nucleus contains two nucleoli (the dark spots) and the chromatin is not condensed. The small circle attached to the upper right surface of the nucleus is the *centrosome*. In (B) the centrosome divides and the halves separate, generating a set of *microtubules* between them; this is the beginning of the *mitotic spindle*. The chromosomes begin to condense. (C) shows a continuation of this process; the nucleoli have become faint. In (D) the nucleoli and nuclear envelope have disappeared, the halves of the centrosome have moved to the poles of the nuclear region, and the spindle extends between them. The daughter chromosomes are aligned along the middle of the spindle. In (E) the spindle is contracting and pulling the daughter chromosomes apart, and in (F) two nuclei have begun to reform around each pole of the spindle; each daughter nucleus contains a full complement of chromosomes. The constriction in the middle of the cell will continue until two separate daughter cells are formed. The technical terms for these stages of mitosis are *prophase* (B, C), *metaphase* (D), *anaphase* (E) and *telophase* (F)

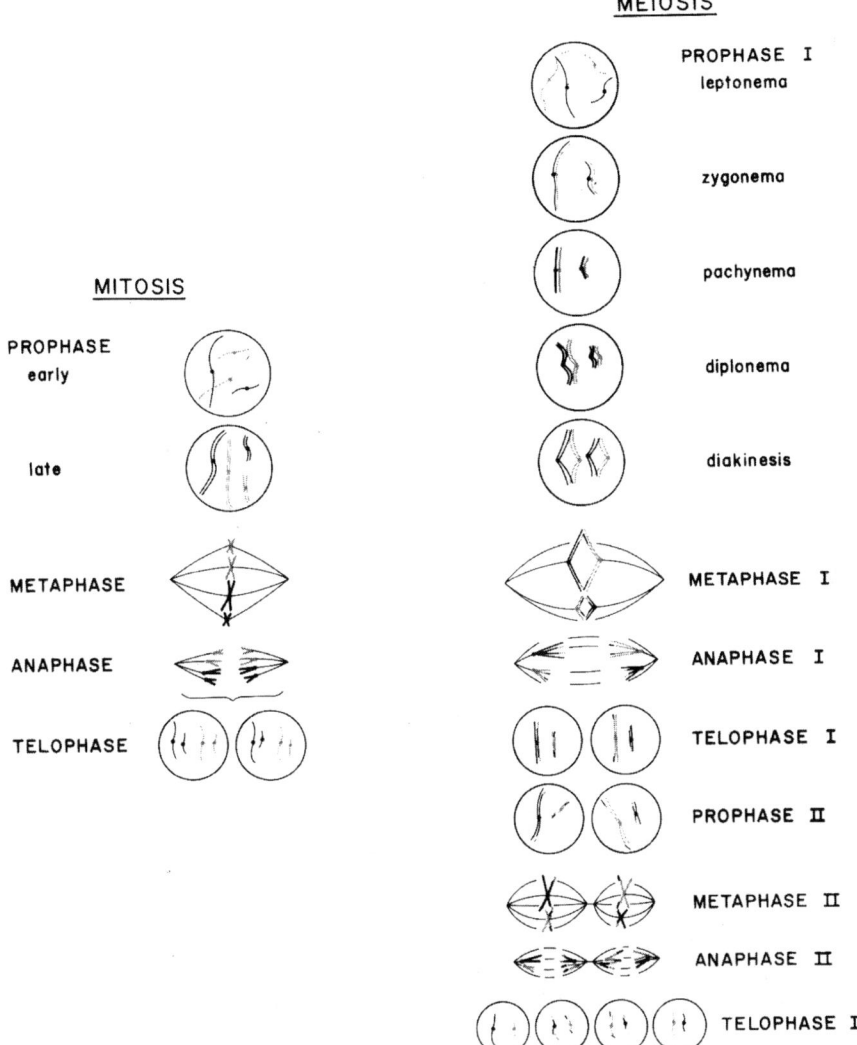

Fig. 10.4 Meiosis and mitosis. Schematic sketch contrasting the 'ordinary' cell division process, mitosis (shown in more detail in Fig. 10.3), with the specialised division process that produces gametes, meiosis. Mitosis ensures that both daughter cells have all the same *paired* chromosome complement as the parent cell. Meiosis is a more complex process in which many distinct stages can be identified. It generates cells with only half the number of chromosomes (*haploid*) that are found in most of the somatic cells of the same organism (*diploid*). The identification of chromosome pairs in diploid cells is discussed further in Chapter 13. Reproduced by permission of Saunders College Publishing from p. 15 of Browder LW (1984) Developmental Biology. Holt-Saunders, Philadelphia, PA. Copyright Harcourt Brace College Publishers

again and appeared more or less anatomically normal. Thus, the mosaic hypothesis was corroborated up to a point, but was evidently too simple to account for all the observations. Roux played down the observations that were inconsistent with his views and attempted to explain them *ad hoc*. That is a common (and understandable) practice among scientists who are committed to a particular stance (Fig. 10.5).

Inspired by his mentor, Haeckel, Roux extrapolated from his results to a 'Darwinian' description of embryo development. He envisaged the process as a fight among the cells and parts of organisms, the stronger cells leaving more progeny, becoming more prevalent and accumulating more nutrients. This presumed that all the cells constituting an embryo behave as independent organisms, which is plainly false.[15] In other respects, however, Roux's contributions were positive and in harmony with the new discoveries in cell biology. In 1883, for example, he gave the first complete biological interpretation of mitosis: the process ensures the correct uniform division of the nuclear substance into two equal and identical halves, implying that the number of chromosomes remains constant. Four years later, he

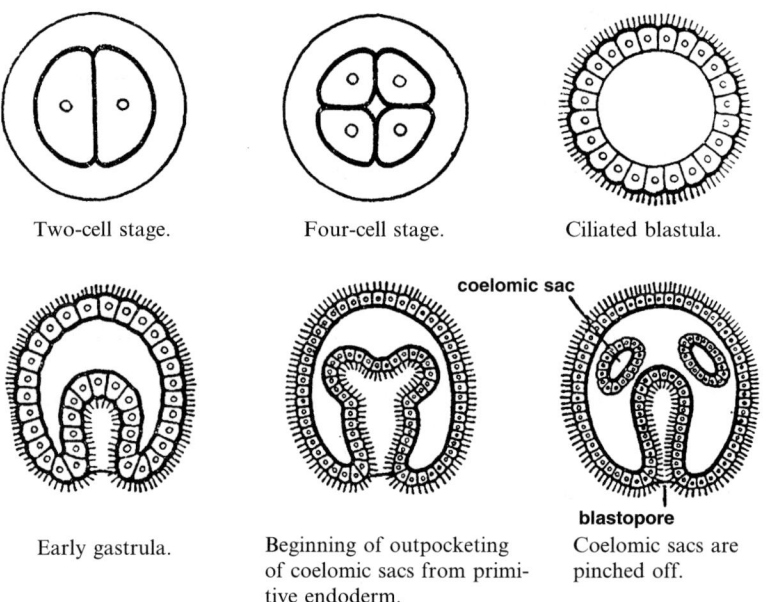

Two-cell stage.   Four-cell stage.   Ciliated blastula.

Early gastrula.   Beginning of outpocketing   Coelomic sacs are
                       of coelomic sacs from primi-   pinched off.
                       tive endoderm.

Fig. 10.5 Early stages of development of a starfish embryo. The zygote (fertilised egg) is cleaved successively to form 2, 4, 8 ... cells until a hollow sphere, the blastula, is formed. In the starfish the cells of the blastula have cilia. In the next stage of development, the gastrula, the three germ layers (ectoderm, mesoderm and endoderm) start to become apparent and from these the tissues of the adult gradually arise. The early stages of embryo development are very similar in most animals

---

[15] Nevertheless, a respectable view of mammalian brain development holds that neurites compete with each other during growth and only the first of each group to reach the target cell survives. This is known as 'neural Darwinism'.

suggested that the linear arrangements of chromosomes are transmitted equally to both daughter cells (gametes) at meiosis.

While Roux's experiments led him to support the mosaic development hypothesis (latter-day preformationism), the studies by Hans Driesch (1867–1941) implied the opposite position (latter-day epigenesis): the hypothesis of *regulative development*. Driesch separated the cells of very early sea urchin embryos by shaking them vigorously in sea water and then cultivating them separately. Each separate cell matured into a normal, if rather small, sea urchin. He was puzzled and disturbed by this result, which he considered a 'backward step' in understanding, but the findings were reproducible. He was forced to conclude that for sea urchins at least, the mosaic development hypothesis was wrong. Although each cell in an undisturbed embryo is 'directed' to forming a particular part of the mature animal, this constraint is removed if the cell is separated. Thus, the whole embryo seems to exert some kind of harmonious control over its parts. Driesch could not reconcile that with any mechanistic account of development. He therefore believed that his results had confirmed Aristotle and attributed his findings to entelechy. Later, he abandoned embryology and became a philosopher, fascinated by vitalism and its relevance to 'parapsychology'.

It gradually became apparent that embryos of some species show relatively 'mosaic' development while others show relatively 'regulative' development. Studies such as those by the pioneering American cell biologist Edmund Wilson (1856–1939) showed that neither account was wholly satisfactory for any organism.[16] The zygote necessarily has the potential to form the whole of the new individual, but the potential of each individual cell becomes more restricted as development proceeds. The rate at which this restriction of potential occurs varies among species, being slow in sea urchins (hence Driesch's results) but relatively quick in amphibians (hence Roux's results). Wilson and his colleagues, Charles Whitman, Edwin Conklin and others, refined this account by studying specific *cell lineages* in the developing embryo, using naturally-occurring pigments to trace the descendents of each cell. Later, vital dyes were employed for this purpose. These researchers were led to distinguish between the *fate* of a cell (the structures that it generates in the undisturbed embryo) and its *competence* (i.e. the structures that it *can* generate if transplanted to a different part of the developing embryo). Fate and competence are identical in any part of an embryo that has become *irreversibly determined*.

Wilson's studies also showed that the cytoplasm of the fertilised ovum has an ordered pattern that directs subsequent development, as His had hypothesised, but this ordering is determined by the nucleus. He wrote[17]:

---

[16] Wilson was a highly original and influential biologist, whose colleagues and students were responsible for many important advances in embryology and genetics. He himself showed by comparative embryological studies that molluscs, flatworms and annelids had a common ancestor. He also discovered the XX/XY system of chromosomal sex determination, though the same discovery was made independently and more or less concurrently by Nellie Stevens.

[17] (1902) Experimental studies on germinal localization. I: the germ-regions in the egg of *Dentalium. J Exp Zool* 1:1–72.

> It therefore appears… probable, that every cytoplasmic differentiation, whether manifested earlier or later, has been determined by a process in which the nucleus is directly concerned, and that the regional specifications of the egg-substance are all essentially of secondary origin.

Like many of Wilson's insights, this was prophetic. Nowadays, we accept that components of the cytoplasm control gene expression in the nucleus, but it is the expression of particular genes that puts those components in place to begin with.

Fate and Competence

Methodological and conceptual innovations during the 20th century made the ideal of 'mechanical epigenesis' envisioned by Maupertuis and Buffon into reality. Two important new concepts that clarified the relationship between fate and competence were introduced by experimental embryologists in the first half of the century:

- During the 1920s, Hans Spemann (1869–1941) and his students found that some parts of an embryo (*organisers*) implement the fates of other parts (*reacting tissue*) at the gastrula and later stages. When the reacting tissue has the appropriate competence and is in contact with the right organiser at the right time, it begins to develop in a programmed way to form more elaborate structures. Correctly timed contact with the organiser determines the fate of an embryonic tissue, depending on its competence.
- During the 1930s, Sven Hörstadius (1898–1996) established that organisers produce chemical substances that affect the development of embryonic structures. These substances (*morphogens*) form concentration gradients between the poles of the blastula. Reacting tissues respond to particular concentrations of morphogens, and these responses determine the fates of their component cells.

The identities of many morphogens and organisers remain elusive. However, these two concepts have yielded an epigenetic model of embryo development that does not have recourse to entelechy. No matter what a morphogen or an organiser *is*, we are now sure that it alters the target cell or reacting tissue, i.e. induces *differentiation*, by *changing the subset of genes that are expressed* in that cell or tissue. The whole set of genes characteristic of the species is present in the zygote, arrayed along the chromosomes. The new individual is 'preformed' only in this highly abstract sense. The embryo then develops by three processes: cell division, cell migration and differentiation. Each of these processes can now be explained mechanistically, thanks to our understanding of the cell cycle, our knowledge of the control of gene expression, and the concepts of organisers and morphogens.

None of this denies the *fact* that embryo development is purposive or goal-directed. However, that purposiveness has now become explicable in *scientific* terms. Embryology, the branch of biology that gave rise to the notion of entelechy at the hands of Aristotle and became a topic of controversy after the Scientific Revolution, no longer demands a perspective that is either unsatisfactorily 'mechanistic' (classical preformationism) or fundamentally teleological (classical epigenesis). The magnitude of this achievement should not be underestimated. Molecular biology has made and continues to make a very important contribution, but the crucial advances were the two great 19th century theories on which the entire scientific status of modern biology depends: cell theory and evolutionary theory. As we shall see in the next chapter, these two pillars of our understanding were equally crucial in overturning another staple of Aristotelianism, spontaneous generation.

Chapter 11
Spontaneous Generation

Everyday observations suggest that certain organisms are spontaneously generated: mushrooms appear suddenly on lawns and manure heaps, the still pond becomes coated with green slime, parasites manifest themselves in animal guts. Spontaneous generation was therefore a widespread belief in most cultures, and it persisted in Europe until the 19th century. We now explain such phenomena in terms of eggs and spores too small to be seen with the naked eye, but such eggs and spores were matters of speculation until microscopes developed into useful instruments.

'Spontaneous generation' is really two hypotheses, not one. *Abiogenesis* is the alleged production of life from matter that is not and never has been living. *Heterogenesis* is the alleged production of (new) life from matter that was once living or associated with life but is now dead or detached from life, such as corpses or manure. We will need to distinguish between abiogenesis and heterogenesis[1] in some parts of this chapter, but mostly we can ignore the distinction and merely write 'spontaneous generation'.

What *sorts* of organisms might be generated spontaneously? Most authors in Greco-Roman and mediaeval times, following Aristotle, believed in the spontaneous generation of 'lower' plants and animals; there was more doubt about vertebrates. The Earth, if not the cosmos as a whole, was seen as akin to an organism: 'Mother Earth' was in effect alive. When almost the whole human population lived in continual contact with death and birth, the idea seemed natural. This was the context in which alchemy flourished, seeking transformations of substances by conveying the characteristic 'living' spirit of one to another. Spontaneous generation was part of that ethos.

Many Scholastic writers held that the embryonic development of 'higher' animals, certainly of humans, was a progression from lower forms (with only nutritive or vegetative souls) to higher forms (which meant, in the case of humans, acquisition of a rational soul). Organisms with only nutritive souls, or very elementary

---

[1] This distinction would have made no sense prior to the Scientific Revolution; the essentially 'organic' conception of all matter implied that the potential for life was omnipresent in the world. Thus, mediaeval scholars would have found it hard *not* to believe in spontaneous generation even without Aristotle's authority on the subject.

P.S. Agutter, D.N. Wheatley, *Thinking about Life*,
© Springer Science + Business Media B.V. 2008

perceptual souls, did not develop in this way; they were formed directly from non-living matter. Aquinas wrote[2]:

> The higher a form is in the scale of being, the more intermediate forms and intermediate generations must be passed through before that finally perfect form is reached. Therefore in the generation of animals and man, these having the most perfect forms, there occur many intermediate forms and generations, and consequently destructions, because the generation of one is the destruction of another.

In an earlier passage in the same work, Aquinas described a Neoplatonist 'chain of being', explicitly related to Aristotle's 'ladder of nature':

> A wonderful chain of beings is revealed to our study. The lowest member of the higher genus is always found to border close upon the highest member of the lower genus. Thus some of the lowest members of the genus of animals attain to little beyond the life of plants; certain shell-fish, for example, have only the sense of touch, and are attached to the ground like plants.

However, Aquinas did not suggest evolution, as many of his Muslim predecessors had done (Chapter 4). He pictured the series *inanimate matter, plant, animal, man, angel* as a continuum, but it was a *static* series not an evolving one (Fig. 4.1).

This way of thinking persisted into the 17th century. For example, Helmont (Chapter 8) believed that molluscs, frogs and even mice could be spontaneously generated. Indeed, he wrote a recipe for the spontaneous generation of mice. But the rising empiricism of the 17th century cast increasing doubt on the notion. By the 1670s, spontaneous generation was widely accepted in respect of only three groups of organisms: insects, particularly flies; intestinal parasites; and microorganisms, recently revealed to the Royal Society by van Leeuwenhoek's microscopes.

Harvey seems to have had mixed opinions. His **omne vivum ex ovo** (Chapter 7) contradicts spontaneous generation, but elsewhere he refers to the heterogenetic origins of parasites such as tapeworms. His surviving works do not tell us how he resolved these apparently conflicting views, but the mere fact that so eminent a figure in the history of medicine had no single clear opinion about spontaneous generation indicates the profound hold that this Aristotelian notion exerted on the 17th century mind.

Redi's Experiments: Insects Are Not Spontaneously Generated

At the end of the 17th century the spontaneous generation of insects was investigated experimentally for the first time. Francesco Redi (1626–1698), a poet, Classical scholar and physician, served at the Medici court of Florence. He had

---

[2] (*Contra Gentiles*, ii. 89; *Of God and His Creatures.*) The Platonic-Augustinian conception of the great chain of being was definitely static: each link in the chain was occupied by the one uniquely appropriate species; and although the links were juxtaposed, nothing could move from its God-given position (*Natura non facit saltum* = nature does not make a jump). The great chain of being profoundly influenced western thought for centuries.

already proved himself a precise and meticulous observer and reasoner in his study of snake bites. He was also well versed in the works of Harvey and the new spirit of natural philosophy. In 1668 he contrived a simple masterpiece of experiment design. According to his own account he was inspired to undertake these experiments by the scene in the *Iliad* in which Thetis promises to keep the body of Patroclus from rotting by protecting it from flies.[3] Were maggots generated spontaneously from dead flesh, wondered Redi, or were they, as Homer seemed to suggest, produced by flies?

Essentially, Redi's experiments involved putting organic matter (various kinds of meat, slices of dead snake, fish, vegetables) into three identical flasks. One flask was sealed tightly with paper, the second was covered with fly-proof muslin, and the third was left open. The experiments were conducted during the summer when flies were plentiful. The results were unequivocal: irrespective of whether air could be exchanged between the flask and its surroundings (the point of the muslin-sealed flask), maggots only appeared in the *open* flask, which flies could be seen entering and leaving. Importantly, the maggots belonged to the species of flies that had entered the flask, no matter what kind of organic matter the flask contained (Fig. 11.1).

Redi went on to disprove the spontaneous generation of bees from the viscera of bulls (a widespread Classical belief), of frogs from mud and of lice from decaying fowls. He speculated that all living things are the progeny of existing plants and

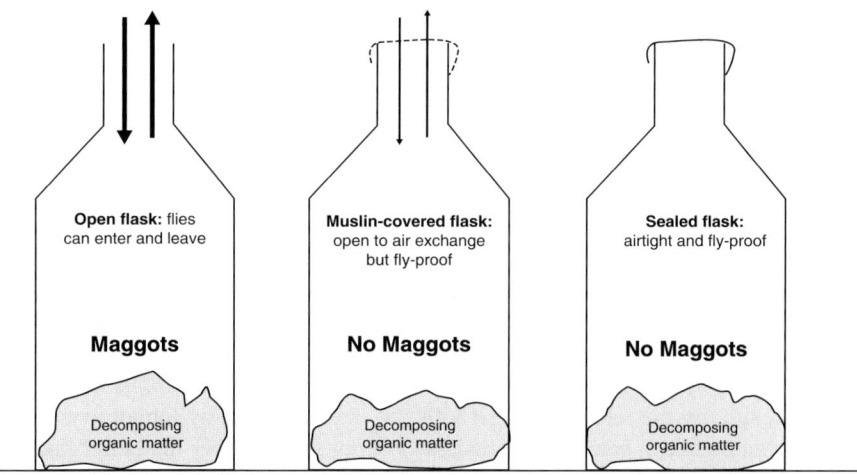

Fig. 11.1 Redi's experiment. Three flasks containing decomposing organic matter are compared. One flask is open, the second is covered with muslin and the third is sealed. Air can enter and leave the first two flasks but not the third. Flies can enter and leave only the first flask, not the other two. Maggots appear only in the first, open, flask. Therefore, maggots form on the organic matter only if it is accessible to flies

---

[3] Few other experiments in the history of science have been inspired by Classical literature!

animals, maintaining the integrity of species. This clear statement of the continuity of life, decidedly ahead of its time, inspired further studies. By the end of the 17th century it was generally agreed that all insects, and external parasites such as mites, reproduced by laying eggs; they were not the products of spontaneous generation. The hypothesis that insects and other arthropods are spontaneously generated had been refuted experimentally, primarily by Redi, and was now pronounced dead. But such experiments were only conceived and executed because the dominant world-view in western Europe was no longer 'mediaeval'. The Baconian approach to knowledge had become accepted, and the alchemical notion that matter was animated by spirits was in retreat as chemistry began to emerge (Chapter 8).

Intestinal Parasites

Redi made a thorough observational study of 'animals found within the bodies of other animals', but although he hinted at life-cycles, he (like Harvey, apparently) could not exclude spontaneous generation in these cases. Indeed, spontaneous generation remained the favoured explanation for intestinal parasites until well into the 19th century.

Using the improved microscopes of the 1830s, Christian Ehrenberg (Chapter 9) described intestinal worms in minute anatomical detail and drew attention to their complex internal forms and copious egg production. Notwithstanding his belief in *Naturphilosophie* and the ladder of nature, he inferred that these organisms were unlikely to arise by spontaneous generation. Most of his contemporaries disagreed.

Alternation of generations in animals (odd-numbered generations have one anatomical form and way of life, even-numbered generations a radically different one) had been discovered in several species of salps by Adelbert von Chamisso (1781–1838) shortly after the Napoleonic Wars. It had subsequently been noted in other species but was regarded as a curiosity until the 1840s, when Japetus Steenstrup (1813–1897) discovered that it was widespread among animals (as well as among plants) and, crucially, that it was universal among intestinal worms. Steenstrup showed that intestinal worms produced eggs, which hatched into free-swimming embryos, which in turn infected an intermediate host. Transfer from intermediate to primary hosts, including humans, was demonstrated early in the 1850s (Fig. 11.2).

As a result, by the middle of the 1850s, no one believed any longer in the spontaneous generation of intestinal parasites. Twenty years earlier the belief had been more or less universal.

Almost two centuries had elapsed between the refutation of spontaneous generation in insects by Redi and his successors, and the refutation of spontaneous generation of intestinal parasites by Steenstrup and his successors.[4] The durability of this

---

[4] Prominent among these was Gottlob Friedrich Heinrich Küchenmeister (1821–1890), who identified the trichinosis parasite. Besides his skill as a parasitologist, Küchenmeister was an eminent theologian and Hebrew scholar. He interpreted the 'fiery serpents' that afflicted the wandering Israelites after their departure from Egypt (Num. 21:6) as Guinea worms, probably correctly.

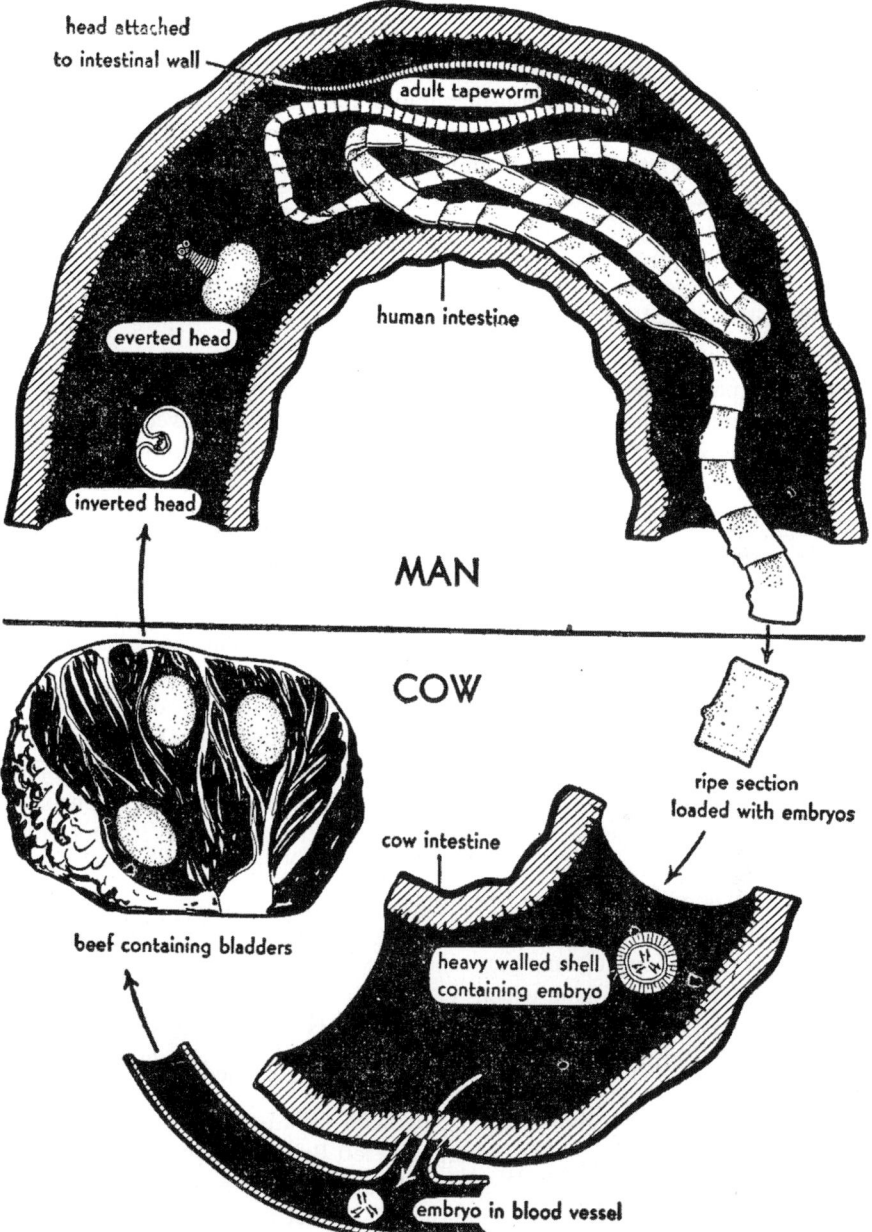

head attached
to intestinal wall

adult tapeworm

everted head

human intestine

inverted head

MAN

COW

ripe section
loaded with embryos

cow intestine

heavy walled shell
containing embryo

beef containing bladders

embryo in blood vessel

Fig. 11.2 Alternation of generations. This diagram illustrates the transmission of the tapeworm *Taenia saginata* between humans and cattle. Segments of the adult tapeworm in the human intestine are shed when mature and leave the body via the faeces. These segments contain microscopic embryos surrounded by a strong shell equipped with hooks. When cattle eat grass containing these segments, the embryos are liberated, attach themselves to the intestinal wall and pass through into the blood stream, where they lose their shells, penetrate the muscle and develop into 'bladder' forms. The bladders are ingested by humans eating undercooked beef and develop in the intestine into adult tapeworms

Aristotelian idea suggests that it was the 'default option': unless continuity of life could be demonstrated, spontaneous generation remained plausible.

Demonstrating continuity of life in microorganisms was particularly difficult, so the debate about spontaneous generation in this case was more prolonged, complex and multifaceted.

Spallanzani Versus Needham

By the early part of the 18th century the Lockean spirit of the Enlightenment was gripping the intelligentsia of Western Europe. Newton had revived the moribund Royal Society, which became the flagship of 'experimental philosophy'. Redi's experiments provoked further study, and van Leeuwenhoek's discoveries cried out for investigation. Where did microorganisms come from?

Leeuwenhoek himself had been dismissive of spontaneous generation, but his attempts to prove that his microorganisms ('animalcules') were not spontaneously generated yielded equivocal results. He designed experiments similar to Redi's, but organisms appeared even in his sealed vessels – anaerobic bacteria, perhaps? During the 1740s it was shown that polyps reproduce by binary fission.[5] If binary fission also occurred among microorganisms, then microorganisms might show continuity of life. This speculation attracted thinkers who sought theistic explanations for biological phenomena: Divine Creation entailed a time at the beginning of history when all species were created, and as Swammerdam had observed (Chapter 6), this belief was incompatible with spontaneous generation. However, spontaneous generation remained the favoured explanation among the many 18th century French intellectuals who rejected Divine Creation.

That group included the century's greatest contributor to natural history, Buffon, though his version of spontaneous generation was subtle. His *Histoire Naturelle* envisaged a kind of heterogenesis in which 'vital atoms' from the remains of dead organisms re-organised themselves into new living forms. However, in accordance with the spirit of the age, Buffon was not willing to rely on speculation. He wished to demonstrate the spontaneous generation of van Leeuwenhoek's animalcules experimentally.

For this purpose, he invited John Needham (1713–1781) to Paris in 1748. Needham, a Catholic priest whose family had left Britain to avoid religious persecution, lived most of his life in Belgium, though he had become a Fellow of the Royal Society in 1747. Needham held, not altogether unjustly, that opponents of the spontaneous generation of microorganisms were merely arguing by analogy from Redi's experiments. He did not quite agree with Buffon; he believed in an omnipresent 'vegetative force' that could give life to otherwise inanimate matter, almost an alchemical notion. Nevertheless he conducted the requisite experiments.

---

[5] The simplest form of reproduction, found in many single-celled organisms including all bacteria: the cell divides into two cells, each a new organism.

He 'violently' heated almond seed infusions, mutton gravy and other media in phials and then stopped the phials with corks.[6] The media became infected. Buffon was satisfied that spontaneous generation had been demonstrated experimentally and he duly described the results in *Histoire Naturelle*.

Lazzaro Spallanzani (1729–1799), also a Catholic priest with a remarkable aptitude for experimental design,[7] replicated Needham's work during the 1760s. His experiments were designed and conducted so meticulously that few could match his technical standards. Spallanzani believed that Needham had not heated the phials sufficiently to kill 'animalcules' already present in the media, and that he had used porous corks. In his own experiments he boiled phials containing a wide range of organic materials for various times up to two hours, then sealed them hermetically. He used metal as well as glass containers, excluding the possibility that the material of the phials could have affected his findings. In most of his media there was no microorganism growth.[8]

Needham responded by saying that the prolonged boiling of Spallanzani's media had destroyed the vegetative force necessary for life, and that the excessive heat had destroyed the elasticity of the air in the phials. In reply, Spallanzani deliberately allowed his boiled media to become infected and showed that prolonged boiling encouraged rather than inhibited growth. Also, he drew out the necks of his phials into capillaries, allowing the air pressure to equalise with the external atmosphere before sealing them hermetically. These new experiments confounded Needham's objections, though microorganisms inevitably grew in a few of Spallanzani's phials. Even the best experimentalists cannot attain perfection all the time. Spallanzani concluded that some animalcules had very heat-resistant 'seeds'. Further experiments showed that they were also resistant to other extreme conditions such as freezing.

Spallanzani's experiments were models of precision and thoroughness but they did not persuade his contemporaries. Other investigators tried to replicate them but, not surprisingly, they found Needham's results easier to obtain. Also, established opinion was against Spallanzani; Buffon's authority carried weight.[9] Of course, the

---

[6] Needham J (1748) A summary of some late observations upon the generation, composition and decomposition of animal and vegetable substances. *Phil Trans Roy Soc* 45:615–666.

[7] This indefatigable worker deserves to be better known. He discovered echolocation in bats; another fine piece of observational and experimental science, though cruel by our standards since it involved blinding, deafening and otherwise mutilating some of the bats that inhabited the bell-tower of his church. Echolocation was not accepted as an animal navigation mechanism until after the invention of sonar, 150 years later, another example of a technological innovation giving rise to a scientific idea. Spallanzani was then 'rediscovered' retrospectively. It has been suggested that his alleged eccentricity, together with his study of echolocation, gave rise to the expression 'bats in the belfry'. He also proved that semen is necessary for fertilisation in amphibians; these experiments involved equipping male frogs with miniature trousers immediately prior to copulation.

[8] Prescott F (1930) Spallanzani on spontaneous generation and digestion. *Proc Roy Soc Med* 23:495–510.

[9] So much for the 'Age of Reason'! According to Bacon and Locke, human authority was supposed to count for much less than the evidence of the senses and the results of experiments. However, human nature had not (and has not) changed since mediaeval times.

general hypothesis of spontaneous generation cannot be refuted experimentally, because the number of conditions under which life *might* conceivably arise is indefinitely large, and no one can conduct an indefinitely large number of experiments. However, Spallanzani's aim was specific: he only set out to show that microorganisms did not originate as Needham had claimed. Buffon, in contrast, considered the hypothesis globally and believed that Needham's results were illustrative of a general process.

During the remainder of the 18th century, several natural philosophers tried to achieve spontaneous generation under experimental conditions; Erasmus Darwin's attempts to 'produce life in a glass jar' were much discussed. Their recurrent failures did not discourage them. Clearly, Spallanzani's high-quality experiments had less impact on the consensus opinion than Needham's less well-executed study and Buffon's authority.

The Needham-Spallanzani debate had an unexpected by-product in the early 19th century: the invention of food canning. Food was heated in a container that was subsequently sealed with an airtight seal, and this became accepted in France as a method of food preservation and hygiene. In effect, it was a commercial application of Spallanzani's experiment. Effective technology was derived from a scientific study that the consensus had, at least for the time being, rejected.

Changing Fashions

Prevailing opinion turned against spontaneous generation as the 19th century advanced. A salient contribution to this shift in the consensus was the proof of alternation of generations in intestinal worms, as mentioned above. However, even before Steenstrup's work in the 1840s, a sceptical trend had become apparent in France, and more notably in Germany, where the idea of spontaneous generation had previously found favour.

This change in attitude may have been encouraged by the political trend against Republicanism, which had become linked with belief in spontaneous generation (see Chapter 12). In addition, the influence of the *philosophes* of the Enlightenment, including Buffon, had lessened. Widespread suspicion of *Naturphilosophie* among Kantians and positivists was another significant contribution.[10] But perhaps the most important factor was the emergence of cell theory. If all organisms consist of one or more cells, and if cells only arise from other cells, then spontaneous generation is logically excluded.

---

[10] Schelling (Chapter 9) believed that Nature was informed by an innate organising principle that struggles towards self-consciousness. This was later developed in the doctrine of the *Will*, due to Arthur Schopenhauer (1788–1860). Neither Schelling nor Schopenhauer had much to say about spontaneous generation, but their philosophical views were clearly compatible with the hypothesis.

In this climate, several experiments were conducted that challenged the spontaneous generation of microorganisms. These experiments were no better designed or
executed than Spallanzani's but in the changed climate they met with a more positive reception. In 1836, Max Schultze (1815–1874) showed that a heat-sterilised
organic infusion remained sterile if it was fed with air that had been drawn through
concentrated sulphuric acid, which must destroy any airborne organisms. It made
no difference if the infusion was provided with warmth and light. An obvious
objection to this experiment was that volatile contaminants from the sulphuric acid
could have killed growth in the medium. A year or two later, Theodor Schwann
performed several well-designed experiments that included drawing heat-sterilised
air through the sterilised infusion; again, there was no growth in the medium. The
principal objection this time was that the heating had destroyed the 'vital ingredi-
ent' in the incoming air, but Schwann was unrepentant. His cell theory[11] entailed
the continuity of life; spontaneous generation was inconsistent with it.

Definitive experimental evidence against the spontaneous generation of microorganisms in organic infusions was obtained by Heinrich Schröder (1810–1885)
and Theodor von Dusch (1824–1890) in 1854. They filtered the air entering their
infusions through a cotton wool plug; no growth occurred in most of their sterilised
media. This experimental design evaded the objection that the air had been damaged by heating or by passing through a noxious substance such as sulphuric acid.
There was growth in some media, but such growth did not necessarily require air.
Schröder and von Dusch had presumably cultured anaerobic organisms.

Later, Schröder showed that by pre-heating the infusions at 130°C and 3 atmos-
pheres pressure, growth could be prevented in *all* experiments. He concluded that
microorganism growth resulted from seeds or germs that came from the air or were
present in the medium at the outset. Techniques now universal in microbiology
laboratories had been invented: the use of cotton-wool air filters, and the autoclav-
ing of culture media. Schröder, however, was careful not to claim that he had
proved the *universal* non-existence of spontaneous generation. The hypothesis still
remained viable – just.

Pasteur Versus Pouchet

Pasteur discovered stereoisomerism (the 'handedness' of molecules; Chapter 8),
established the biological basis of fermentation and the germ theory of disease, and
introduced vaccination to prevent rabies. His experimental disproof of spontaneous
generation ranks alongside those great achievements. Pasteur's study on this subject
is rightly considered definitive and is summarised in many modern textbooks of

---

[11] Schwann's original idea was that cells were formed from amorphous material by a process akin
to crystallisation, and this need not entail the continuity of life. However, that idea was soon
replaced by the Raspail-Virchow conception that cells arise *only* from other cells (Chapter 9).

microbiology. The original swan-necked flasks that he made and used in the crucial experiments still remain uncontaminated after a century and a half. However, the original literature suggests that the events surrounding his work were not as straightforward as is usually supposed. Besides being an outstanding scientist, Pasteur was a political manipulator.[12]

In Chapter 8 we mentioned his attempts to impose 'handedness' on molecules by subjecting them to strong magnetic or electrical fields, presumably to 'create life', as Erasmus Darwin and his contemporaries had sought to do. Pasteur was seldom reticent about his work, but he kept his notes on these experiments secret and mentioned them only in private letters to his uncle. Perhaps he felt that they were incompatible with his political and religious leanings: he was a personal friend of the Emperor Louis Napoleon and a fervent supporter of his right-wing, pro-Catholic, anti-Republican government, so he rejected Lamarck's ideas of species transformation and spontaneous generation.[13] To be consistent, he should presumably have repudiated any attempt to 'create life' in the laboratory.

In 1859 an elder statesman of the medical and biological fraternity, Félix-Archimède Pouchet (1800–1872), published a now almost forgotten book entitled *Hétérogenie*. Pouchet was Lamarkian by inclination, though he was a staunch Catholic. He argued a new version of the heterogenesis hypothesis, attempting to reconcile it with the *omne vivum ex ovo* principle. According to Pouchet, eggs rather than mature organisms can arise spontaneously from dead organic matter. Almost half of the text of *Hétérogenie* is devoted to showing that spontaneous generation is consistent with catastrophist rather than uniformitarian geology and to reconciling the idea with the Scriptures. Pouchet claimed to have repeated many experiments on spontaneous generation, including Schwann's, and to have obtained microorganism growth. Members of the Académie des Sciences who examined his claims pointed out the need for high sterilising temperatures in such experiments, the rapid proliferation of 'infusoria' (microorganisms), the presence of microbes in air and rainwater, and the accumulated mass of contrary experimental evidence. In short, they were sceptical about Pouchet's claims. But they gave him a hearing.

In experiments such as those conducted by Schwann, some flasks are likely to become infected; the less technically competent you are, the more flasks will be contaminated, so the more 'evidence for spontaneous generation' you will find. It is important to decide which 'rogue' results are scientifically significant and which are simply the consequences of technical error; that is part of the art of science. Pouchet's later studies, if not his earlier ones, suggest that he was not an accomplished experimentalist. Therefore, his results perhaps *did* contradict Schwann's, but only for reasons of poor experimental technique.

---

[12] He certainly had a taste for controversy with political overtones. As a young man he clearly relished his disagreement with Justus von Liebig, the doyen of organic chemists, about fermentation. Pasteur seems to have regarded his victory in this dispute as, among other things, a triumph of French science and culture over German.

[13] We shall discuss this more fully in Chapter 12. The key point is that belief in spontaneous generation had become associated with 'left-wing' Republican politics, particularly in France.

Pasteur submitted his *Mémoire* of 1861 for the Académie's prize for 'an essay throwing light on the question of so-called spontaneous generation'. It was a tendentious document. The literature review was biased and the tone of writing was far from dispassionate. Nevertheless, the quality of the experiments was beyond reproach. Pasteur found that a cotton wool plug through which air was passed a 1 litre per minute collected thousands of microorganisms of different species in 24 hours, directly refuting Pouchet's contention that very few such organisms are airborne. These organisms grew in media that were supplied with heat-sterilised air, showing that they could be sources of contamination. Flasks of media opened to the air became infected, but there was less putrefaction of the medium if the flask was opened after a shower of rain or at high altitudes in the Jura Mountains. Moreover, Pasteur showed in the final chapter of his *Mémoire* that microorganism growth can occur on inorganic media, disproving the contention that pre-formed organic materials are necessary for 'spontaneous generation'. He showed that mercury was an unsatisfactory barrier against airborne infection because dust on its surface contained viable organisms. He described a study of the sterilisation of milk that was, in effect, the invention of pasteurisation. Also, he repeated Schwann's and Schröder's definitive experiments, though without crediting the original (German) authors. Perhaps only the celebrated swan-necked flask experiment was truly original, and in principle even that can be criticised.[14] Nevertheless, the *Mémoire* was compellingly clear and rich in both evidence and reasoning. Pouchet withdrew his own entry for the competition and Pasteur was awarded the prize (Fig. 11.3).

But Pouchet did not concede the argument. When he exposed flasks of sterile media to high mountain air in the Pyrenees, most of them became infected. (That was a valid result. He used hay infusion boiled at 100°C. As Ferdinand Cohn

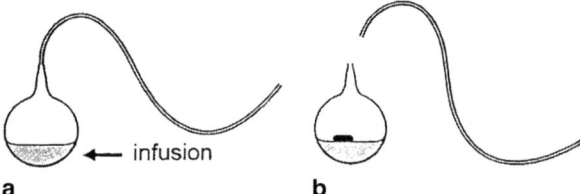

a   b

Fig. 11.3 Pasteur's swan-necked flask. The infusion in the intact flask (A) remains sterile because although air can exchange through the long convoluted neck of the vessel, any bacteria or spores suspended in the air become attached to the glass surface and do not enter the body of the flask. When the neck is broken close to its junction with the flask (B), the infusion is infected. Reproduced from Fig. 10.2 (p. 117) in Harris H (2002) Things Come to Life. Oxford University Press, London. By permission of Oxford University Press

---

[14] Pasteur's flasks were open to the air only through very convoluted narrow necks. When the flasks were sterilised and filled with sterile medium, they remained sterile, even though air could be exchanged through the neck. When the neck was broken the flasks became infected. The point is that airborne microorganisms and spores become trapped in the neck because of its extremely high surface area. But the rate of air-flow by diffusion through the swan neck is infinitesimal, so a proponent of spontaneous generation could claim that there was no growth because insufficient air was supplied. As far as we are aware, however, Pouchet did not offer that criticism.

showed in 1875, hay infusion contains *Bacillus subtilis* spores, which remain viable at that temperature.) As for the presence of organisms in air at lower altitudes, Pouchet tried to prove they were scarcer than Pasteur had shown, but his techniques and arguments were unsatisfactory. For example, he found almost no organisms in the top 5 cm of fallen snow; Pasteur pointed out that there would be more organisms in the *lower* layers of snow, which had fallen earlier.

Pouchet's position was ill-conceived and badly supported, but his cause was not helped by the translation of Darwin's *Origin of Species* into French in 1862. The translator, Clémence Royer (1830–1902), was a radical who affected to see in Darwin's masterpiece a scientific defence of the Lamarckian and Republican position. Her long preface to the translation politicised Darwin in a way that was unacceptable to the French establishment of the time, and further entrenched the linkage between evolution, spontaneous generation and Republicanism in the minds of the intelligentsia.[15]

The debate did not quite die with Pouchet's death in 1871. An English anatomist, H. C. Bastian (1837–1915), continued to argue in favour of spontaneous generation until the end of his life, but he was a lone voice. During the 1880s he came into conflict with the ageing Pasteur, and despite the compelling evidence and arguments of the latter he remained unrepentant. In science as in other areas of culture, there are always champions of lost causes.

Pasteur's rhetoric became passionate. He refused to accept Darwin's theory, which gives a *deductive* reason for rejecting spontaneous generation; but nevertheless he declared that spontaneous generation *cannot* occur anywhere, under any circumstances. In one of his speeches he referred to spontaneous generation as 'That German theory', invoking the intense anti-German sentiments in France after the Franco-Prussian War but ignoring the long-standing German opposition to the hypothesis manifest in the experiments of Schultze, Schwann, Schröder and von Dusch.

Near the end of his life, Pasteur returned in virtual secret to his attempts to achieve abiogenesis in the laboratory. Once again, of course, he failed.

Afterword: The Origin of Life

We do not believe in spontaneous generation because it has been disproved experimentally, it is logically inconsistent with our theory of evolution, it is incompatible with our belief that life is continuous, and there is no conceivable mechanism that could explain it. Yet spontaneous generation (abiogenesis, to be more precise) *did* happen, at least once; life on Earth had a beginning.

---

[15] The fact that Darwin's theory, in contrast to Lamarck's, is logically inconsistent with spontaneous generation apparently eluded both Royer and her conservative contemporaries. Again, we shall explain this point in more detail in the following chapters.

After the key publications by Darwin and Pasteur, which respectively established evolutionary theory and experimentally refuted spontaneous generation, scientists began to ask new questions about the origin of life. In the later years of the 19th century the chemist Svante Arrhenius proposed the notion of 'panspermia' (the seeds of life are present everywhere in the universe and only need a suitable planet on which to 'grow'), and this presaged a trend among physical scientists towards the view that life had an extraterrestrial origin. Early in the 20th century Orgel, Haldane and others, predominantly socialists, began to inquire into 'dialectical' chemical processes that might account for life's terrestrial origins. Orgel's 'coacervates' can be seen as precursors of some modern ideas of abiogenetic macromolecule formation. However, not until the Miller-Urey experiment of the early 1950s was the origin of life made scientifically respectable by an ostensibly appropriate experimental study that yielded interesting if perhaps misleading results (see *About Life*, Chapter 14).

We shall return briefly to this topic in the final chapter of this book, but its importance should not be underestimated. Profoundly difficult though it is to investigate the origin of life scientifically, biology will remain philosophically incomplete until we have consensus about how life on Earth began.

Chapter 12
The Evolution of Darwinism

Like all viable scientific theories, the theory of evolution is continually growing, developing and changing. Here and in the following two chapters we trace its history and show why it is integral to the modern science of biology.

Evolutionary Ideas Prior to 1800

In Chapter 4 we pointed out that mediaeval Europe had inherited two ways of thinking about relationships among species, a Plato-Augustine tradition, which regarded species as fixed and immutable, and an Islamic tradition, which had developed the evolutionary ideas of pre-Classical philosophers such as Empedocles and hinted at a mechanism akin to natural selection. Both traditions evoked the Neoplatonist 'great chain of being', which in turn claimed legitimacy from Aristotle's 'ladder of nature'. Aristotle was not explicit about the matter; he quoted Empedocles with apparent approval, but he believed in spontaneous generation and therefore cannot have considered all species to be products of evolution.

Around the time of the Scientific Revolution, the predominant belief was that species were 'fixed' as God had created them. Some philosophers of that era (e.g. Leibnitz) seemed to promote 'evolutionary' ideas, but they were writing about spiritual rather than material evolution. The preformationist view of embryo development precluded any notion of evolution (Chapter 10). Not until the Enlightenment were 'transformist' ideas proposed. In 1745, Maupertuis offered the first explicit hint of evolution by natural selection since Al-Jahiz in the 9th century:

> [I]n the fortuitous combinations of the productions of Nature, since some must be characterized by a certain relation of fitness and are able to subsist, is it not to be wondered at that this fitness is present in all species currently in existence? Chance, one would say, produced an uncounted multitude of individuals; a few found themselves constructed in such a manner that the parts of the animal were able to satisfy its needs; in another vastly greater number, there was neither fitness nor order: all of these latter have perished.

This coincided with his epistatic view of embryo development. La Mettrie also suggested that organisms might diversify because of heritable random changes. Buffon accepted a more limited 'transformism', believing that the 200 or so then-known species of mammals may have descended from some 38 original forms; those original forms were the results of spontaneous generation, and the 'internal moulds' in which they had been cast limited the amount of variation that were subsequently possible. James Burnett, Lord Monboddo (1714–1799), formulated ideas similar to those of Maupertuis. He reasoned that animals changed their characteristics over time in order to accommodate to the environment. He also considered that humans had evolved from apes, a controversial view in the 1770s. Monboddo was a major influence on Charles Darwin's grandfather, Erasmus (1731–1802).[1]

However, the founder of modern taxonomy, Linnaeus, spoke for the consensus: species were essentially immutable, as decreed by God, though hybrid forms could emerge. Linnaeus's taxonomy superseded Aristotle's and heralded the end of the 'great chain of being' as a major influence on western thought, so it was very influential. Anti-'evolution' though it was, it drew attention to structural similarities among species and encouraged the search for common origins, which partly explains why Erasmus Darwin praised Linnaeus so highly. Followers of Linnaeus did not return the compliment. *Natural Theology* by William Paley (1743–1805), which sought to base biology on a God-centred teleology, was Linnaean. Paley wrote this famous 'argument from design' as an explicit rebuttal of Erasmus Darwin.

The Influence of Geology

Towards the end of the 18th century, efforts were made to establish a scientific geology. These endeavours applied Enlightenment thought and 'experimental philosophy' to yet another aspect of the observable world, but they were motivated by the industrial revolution. Increased mining activity and the need to establish long-range systems of communication and transport (roads, canals and railways) demanded a better understanding of the distributions of rock types and the principles underpinning landform. Within half a century, geologists had established that the Earth was of immense age and that the fossil record probably represented a succession of organisms that had become extinct. Those two notions were essential for the maturation of evolutionary theory.

---

[1] Erasmus Darwin's idea of 'transformism' did not involve natural selection. It was encapsulated in his poems *Zoönomia* and *The Temple of Nature*, the latter being published posthumously in 1803. He believed in a progression of life from microorganisms to civilised humanity, deriving from '… *one living filament, which the great First Cause endued with animality, with the power of acquiring new parts, attended with new propensities …, and thus possessing the faculty of continuing to improve by its own inherent activity, and of delivering down these improvements by generation to its posterity*'. The concept was, in effect, *Naturphilosophie* and the great chain of being with Aristotelian overtones, and it was highly appreciated by Romantics such as Coleridge. This alignment of Erasmus Darwin with the Romantic movement is curious, since he was long associated with the Lunar Society, a group of enthusiasts for experimental philosophy who represented much of the intellectual driving force behind the industrial revolution – just what the Romantics opposed.

Several controversies surrounded the birth of geology. The one most relevant to biology was the dispute between *catastrophists* and *uniformitarians*. Catastrophists held that the Earth's surface was shaped mostly by violent events such as eruptions, earthquakes and devastating floods. Their position appeared consistent with Biblical history. The uniformitarians held that the Earth was formed by slower, gentler processes of erosion, transport and deposition. Their position required the Earth to be much older than Biblical chronology suggested. Thus, the age of the Earth was seriously debated, a new phenomenon in western thought.

Some early uniformitarian writings such as those of the pioneering geologist James Hutton (1726–1797)[2] were explicitly Aristotelian; in contrast, catastrophists such as Cuvier made a serious effort to be 'scientific' from the outset. In 1796, Cuvier showed that mammoths and mastadons were distinct from any modern species, thus proving that species could indeed become extinct. He suggested that a series of catastrophes and subsequent acts of Divine creation had been responsible for eliminating such species and replacing them with others. This view was echoed by 19th century British geologists such as Sedgwick and Buckland. Buckland in particular attempted to equate the last in the series of supposed catastrophes with the Biblical flood.[3]

The miner and land-surveyor William Smith (1769–1839) constructed a geological map of England, ordering the rock strata chronologically by examining the fossils they contained. Smith's work more or less coincided with the stratigraphical study of the Paris basin by Cuvier and Brongniart (Chapter 9). Thanks to these endeavours, the principal features of the geological time-scale had become apparent by the 1840s. The uniformitarian Charles Lyell (1797–1875) published his multi-volume *Principles of Geology* in the early 1830s, establishing the great age of the earth. This work was much admired by Darwin, though Lyell opposed the idea of evolution, even doubting that the fossil record represents a true progression.

Lamarck's Concept of Transformation

Jean-Baptiste de Lamarck (1744–1829) was a disciple of Buffon. He was appointed Curator of the Natural History Museum in Paris after the Revolution and he set himself the colossal task of rationalising the collection of fossils. This was approached by applying Linnaean taxonomic principles to extinct as well as extant organisms, with the emphasis on invertebrates. He came to recognise *lineages* of species, notably animals, some of which had terminated while others seemed continuous

---

[2] The opening pages of Hutton's treatise are explicitly Aristotelian, though the remainder of the work seems to develop in a Baconian manner from careful observation. The second generation of uniformitarians, notably Lyell, whose *Principles of Geology* exerted so marked an influence on Darwin, proceeded more 'scientifically'.

[3] In retrospect, the uniformitarian account seems to accord better with our modern knowledge of evolution. However, we now recognise that the history of life on Earth has indeed been punctuated by a series of major extinctions, much as Cuvier and his successors argued.

with living creatures of today. This led him to propose a process of 'transformation' in his *Philosophie Zoologique* of 1809. The main ideas in Lamarck's transformism were as follows.

- The simplest organisms ('infusoria') originate through spontaneous generation. All subsequent organisms have developed over time from these elementary microscopic forms.
- This 'ascending' or 'progressive' history of life is driven by heat and electricity. Such physical agencies are responsible for the spontaneous generation of micro-organisms and provide the impetus that transmutes them to progressively more complex forms. Lamarck thus implied that living matter has an inherent dynamism that empowers it to generate novel forms and structures, but his argument evoked *physical* forces, not a 'vital force'. Like his mentor, Buffon, he believed that organic *matter* is distinctive.
- The history of life has created fourteen *linear* (as opposed to branching) sequences of forms, culminating in the mammals. This corresponded to the 'natural' linear order of his taxonomic system.
- Major transformations between species may occur as a result of use and disuse of structures.

The notion traditionally ascribed to Lamarck, the inheritance of acquired characteristics, was not particularly significant. Lamarck proposed it as the means by which major animal groups adapt to local environments but not as a primary *cause* of transformation. Belief in the inheritance of acquired characteristics was by no means unique to Lamarck; everyone who accepted transformism or evolution, including Darwin, shared the belief in one form or another. Not until the chromosomal theory of inheritance was established (Chapter 13) could it be seriously challenged (Fig. 12.1).

Influence of Lamarck

Because Cuvier opposed the idea of transformation and was greatly respected, Lamarck's work found little support in France during his lifetime.[4] His main defender was Etienne Geoffroy St Hilaire (Chapter 9), who saw transformation as support for his 'formalist' biology. The debate between Cuvier and Geoffroy was thereafter extended into a disagreement about transformation. After Lamarck's death, Geoffroy wrote[5]:

---

[4] Cuvier's catastrophism was incompatible with Lamarck's views. Cuvier also observed that drawings of animals and animal mummies from Egypt, which were thousands of years old, were identical with modern animals. His reputation as a leading scientist helped keep Lamarckian ideas out of the scientific mainstream.

[5] Geoffroy Saint-Hilaire (1833) Influence du monde ambiant pour modifier les formes animales.

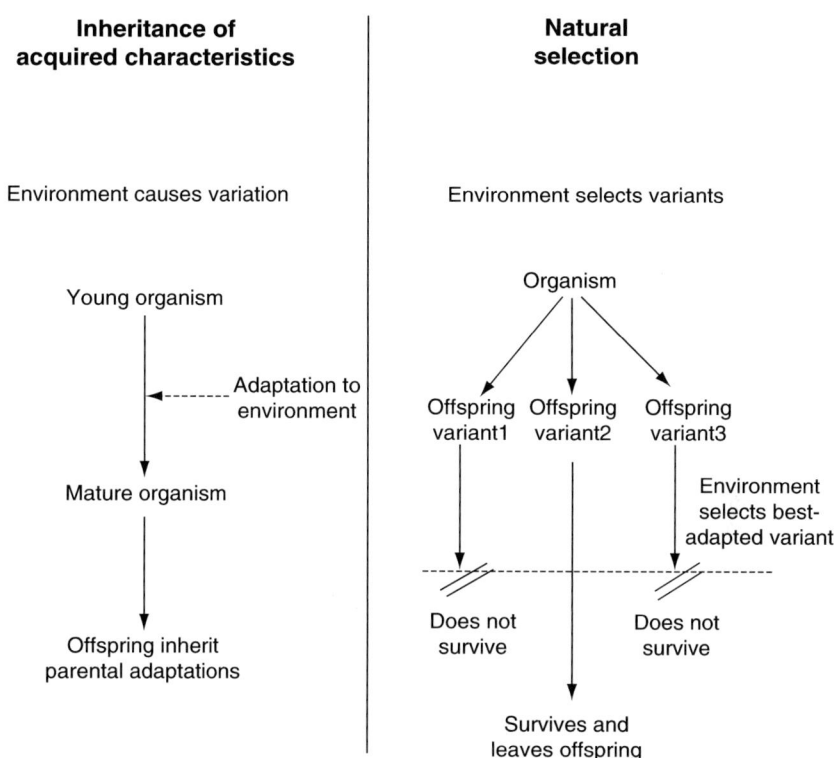

Fig. 12.1 Two views of transformation (evolution). The inheritance of acquired characteristics is a mechanism by which the environment *causes* or 'instructs' heritable changes to occur in an organism. Most of the early believers in 'transformism', including Darwin, accepted that this mechanism plays some part in the process, though it has been associated specifically (and rather inaccurately) with Lamarck. Natural selection is a mechanism by which the environment *eliminates* the less well-adapted variants of a species. Although several writers had envisaged the mechanism, Darwin and Wallace were the first to argue (independently) that it is the main driving force of evolution. It leads to the question: what causes heritable variation?

> 'The external world is all-powerful in alteration of the form of organized bodies... these [modifications] are inherited, and they influence all the rest of the organization of the animal, because if these modifications lead to injurious effects, the animals which exhibit them perish and are replaced by others of a somewhat different form, a form changed so as to be adapted to the new environment.'

This was an explicitly Lamarckian statement and Geoffroy was subsequently praised by Darwin.

Although France in the second quarter of the 19th century was stony ground for the seeds of evolutionary thought, the Lamarckian thesis excited interest in other parts of Europe. In Germany it appealed to the proponents of *Naturphilosophie,* particularly Goethe. Johann Gottfried von Herder (1744–1803) had developed a similar idea of progressive development of species culminating in humanity; like Lamarck, his inspiration was Buffon. In Scotland, Lamarckism was greeted enthusiastically by the comparative anatomist Robert Grant (1793–1874), who saw that

Lamarck's ideas could easily be reconciled with those of Erasmus Darwin. Grant found that his study of homologies supported the idea of common descent of species. An anonymous paper of 1826, probably attributable[6] to the geologist and natural historian Robert Jameson (1774–1854), celebrated Lamarck's explanation of how higher animals *evolved* from simple organisms such as worms. This may have been the first use of the word 'evolve' in its modern sense.

But Lamarck's theory was radically inconsistent with traditional belief and Church teaching. It came to be associated with Republican and sceptical views and was correspondingly opposed by monarchists and Catholics in France and by traditionalists everywhere.[7] English geologists who were influenced by natural theology, such as Buckland and Sedgwick, regularly attacked the ideas of Lamarck and Robert Grant. Concomitantly, evolution (transformism) became a major topic of debate in the clubs and the quarterly journals in England during the first half of the 19th century.

In 1844 the Scottish publisher Robert Chambers (1802–1871) anonymously published a popular book, *Vestiges of the Natural History of Creation*, which was controversial and influential. Chambers proposed that the origin and evolution of life was continuous with the origin and evolution of the solar system, and that the fossil record showed a progressive ascent of animals. The transformations of species were the unfolding of a preordained plan woven into the laws of the universe. The implication that humans were the final step in the ascent of animal life angered conservatives, who believed in the special creation of our species, but the *Vestiges* was a theistic rather than a scientific work. Chambers attributed 'progressiveness' to the universe as a whole, a position adopted by others such as Herbert Spencer (1820–1903). The *Vestiges* intensified the continuing debate about evolution. Its influence partly explains the public reception of the *Origin of Species*.

The eminent anatomist Richard Owen (1804–1892) believed that each species was fixed and unchangeable because it represented an idea ('archetype') in the mind of the Creator. Relationships among species could be revealed by comparing the development of their embryos and by studying the fossil record, but these relationships represented an underlying pattern of divine thought, not a biological process. Progressive creation had led to increasing complexity culminating in the human species. Like Grant, Owen made detailed studies of homologies, which Darwin was to use, and he conceived of *branching* connections among species, not linear ones as Lamarck had supposed. Owen's influence marginalised Grant's position in the scientific community. Darwin's increasing friendship with Owen coincided with a cooling of his relationship with Grant.

---

[6] This has been much debated. Many commentators have presumed that Grant was the author, but a convincing case for Jameson was made by Secord J (1991) Edinburgh Lamarckians: Robert Jameson and Robert E. Grant. *J Hist Biol* 24:1–18.

[7] In England, Lamarckism came to be associated with radical, anti-establishment political movements such as the pioneering socialist movement Chartism, which were socially and politically unpopular (to the extent that their proponents often went to prison), but were major talking-points.

Charles Darwin (1809–1882) and the Natural Selection Model

The Character of Darwin's Theory

Darwin's idea of natural selection developed progressively from its earliest sketch in 1838 until the 1850s, then continued to develop and change through the six editions of the *Origin of Species* published during his lifetime. His later publications elaborated it further. In its most mature form, Darwin's theory differed from previous notions of transformism in four salient ways:

- It concerned only the evolution of *organisms*, not the alleged 'evolution' of the solar system before life, or the origin of life. In this respect it was more limited in scope than the ideas of (e.g.) Buffon, Chambers or Spencer.
- It did not presume that biological evolution is 'progressive', involving the emergence of 'more advanced' organisms from 'more primitive' ones. This distinguished Darwin's theory from (e.g.) Lamarck's.[8]
- It did not suppose that the environment *imposes* evolutionary change on its inhabitants; rather, it held that the inhabitants differ from each other, and some variants are better suited to the environment than others.
- *It was not teleological.* Unlike all other theories of transformation, and in sharp contrast to the tradition of *Naturphilosophie*, the mature[9] form of Darwin's theory did not contain or imply the notions of 'purpose' or 'design'. That is why the mature theory of evolution is the key to solving the problem of purpose in biology, obviating the need for Aristotelian entelechy (see Chapter 15).

Although it is easy to find precedents for the core of Darwin's theory,[10] none of Darwin's predecessors seem to have understood these implications. They assumed either that natural selection was self-evident, or gave abstract arguments for its importance, offering no significant empirical support and no philosophical reflection (Fig. 12.2).

---

[8] *Orthogenesis*, the belief that life has an innate tendency to progress towards ever-greater perfection, retained a significant following: the Russian biologist Leo Berg, the American palaeontologist Henry Osborn and Darwin's German contemporary Heinrich Bronn all maintained some version of this belief.

[9] We stress the *mature* form, first identifiable in the third and fourth editions of the *Origin*. The early editions, and Darwin's notebooks between the 1830s and the 1850s, endowed the theory with either explicit or covert teleological elements suggesting the influence of *Naturphilosophie*.

[10] In the third (1861) and subsequent editions of the *Origin*, Darwin added an introductory essay entitled *An Historical Sketch of the Recent Progress of Opinion on the Origin of Species*, in which he listed these precedents. Some of them were very obscure. They were not known either to Darwin or to Alfred Russel Wallace (1823–1913) when they presented their joint account of natural selection to the Linnaean Society in 1858.

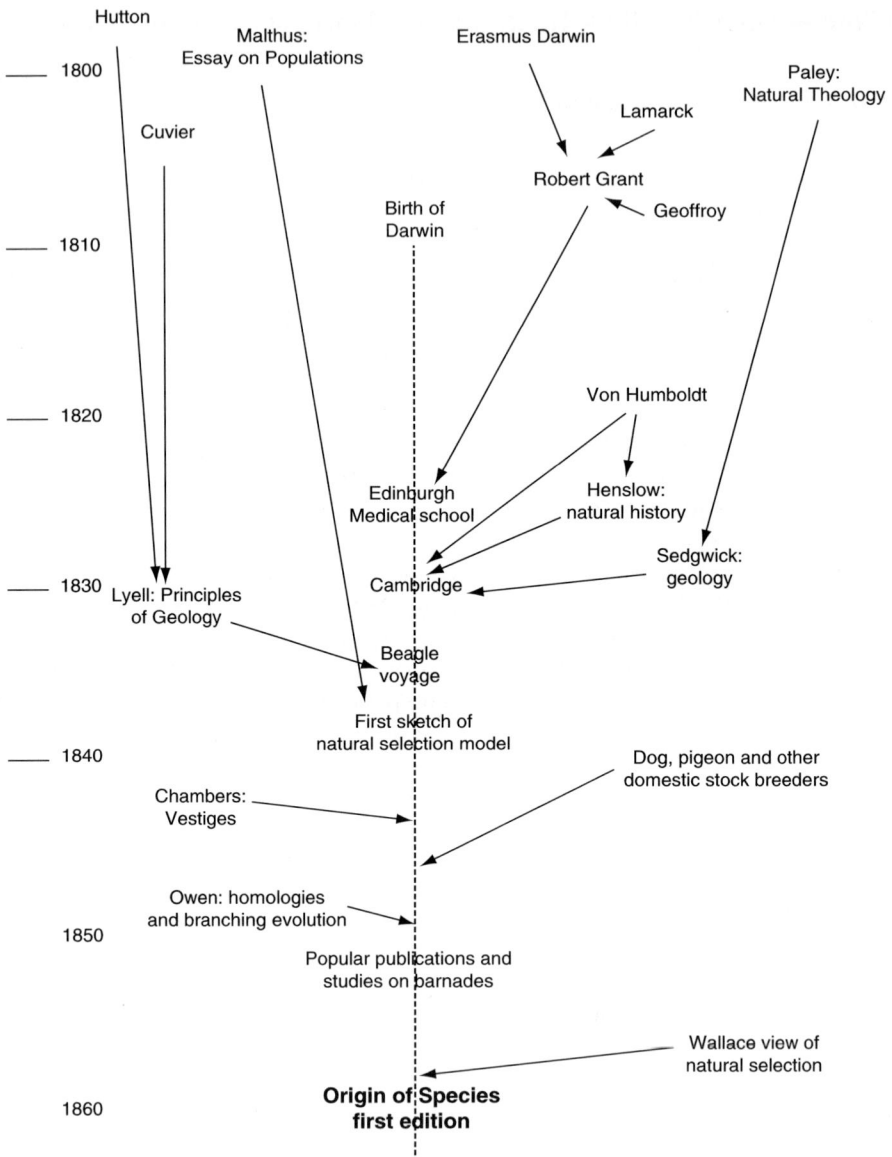

Fig. 12.2 Summary of main influences on Darwin's thoughts about evolution. The vertical dotted line is an approximate time-line of Darwin's life. The approximate chronology of the main influences on his thinking is shown in relation to it.

Formative Influences

Numerous influences contributed to the earliest sketch of Darwin's theory; the extent of each contribution has been much debated:

- The tradition of natural theology represented by William Paley
- The writings of Darwin's grandfather Erasmus and of Lamarck and Geoffroy, which he encountered in Edinburgh (1825–1827) thanks to Robert Grant
- Darwin's coaching in natural history, particularly in botany and entomology, by John Henslow (1795–1861) during his time at Cambridge (1827–1831)
- His field work in geology during the same period, guided by Sedgwick
- The writings on botanical geography, geology and a personal interpretation of *Naturphilosophie* by Alexander von Humboldt (1769–1859), to which Darwin was introduced by Henslow
- The philosophies of science propounded by William Whewell as mentioned in chapter 9 and by John Herschel[11] (1792–1871), and also Mill's version of positivism (see below)
- The copious biogeographical evidence collected during the *Beagle* voyage
- Uniformitarian geology, to which Darwin was converted (again during the *Beagle* voyage) by reading Lyell's *Principles of Geology*
- The sixth edition of the *Essay on Populations* by Thomas Malthus[12] (1766–1834), which Darwin read in the late 1820s. This seems to have been the seed crystal from which the natural selection model took shape. Generalisation of the Malthusian principle to all organisms provided an 'evolutionary dynamic' – though Darwin did not explain *why* living beings tend to reproduce geometrically.

After the initial sketch of the theory in Darwin's 'D notebook' of 1838 and various increasingly detailed elaborations during the following years, two further events were instrumental in the production of the *Origin of Species*:

- The phenomenal popular success of the *Vestiges* by Robert Chambers, which prepared Victorian society for a more plausible theory of evolution.
- The discovery in the late 1850s that Alfred Russel Wallace (1823–1913) had chanced upon the natural selection model of evolution quite independently, again inspired by the *Vestiges,* Malthus's essay, and Wallace's own observations in South America and the Malay Archipelago. Darwin and Wallace presented

---

[11] Herschel's *A Preliminary Discourse on the Study of Natural Philosophy* (1830) propounded a strictly Baconian, inductive approach to science, in line with Mill's positivism, but also taught that the aim of science was to identify laws of universal application, a position close to Whewell's.

[12] Malthus's *Essay*, written in the late 1790s, was a best-seller. Explaining that human populations increased geometrically and the food supply arithmetically, it recommended that the weaker members of the exponentially-growing working class be permitted to starve so that the workforce would become stronger. This argument contributed significantly to the political and economic justification for Britain's poor laws and workhouses. Malthus was not, in fact, original. Benjamin Franklin had stated the principle that population growth outstrips food supply some 20 years before the *Essay* was published. The liberal Franklin, however, used the principle to argue in favour of social welfare – in contrast to Malthus.

the theory jointly to the Linnaean Society in 1858, after which Darwin rushed the *Origin of Species* into print. The book was no more than an abstract of his immense but unpublished compilation of supporting data.

The Argument of the Origin of Species

The *Origin of Species* brought the question of the origins and extinctions of species into the domain of naturalistic, mechanistic explanation. It unified many fields of biology and was scientifically fertile. Darwin was aware of the opposition that his radical account of life would encounter, so he set out to forestall the obvious criticisms. This was achieved as much by the *structure* of the book as by its somewhat summary content. Large bodies of data have the same effect on readers as mathematical equations: they deter them. Darwin wanted to reach the widest possible audience. Nevertheless, the book gives an overall impression of Baconian reasoning.

The book opens quietly and, at first glance, uncontentiously. The first chapter is about the origins of domestic animals (and, implicitly, plants): slight differences among individuals of a species are augmented by selective breeding. The point is illustrated with familiar examples such as pigeons and dogs. Starting the book in this way was an effective rhetorical strategy, but it also furnished Darwin with a way of visualising the central theme, natural selection, which he deferred until Chapter 4.

Chapter 2 tackles the crucial concept of 'species'. Darwin uses the existence of variation, established in Chapter 1, to undermine the Linnaean basis of classification in terms of 'essential defining properties'.[13] Buffon, the most vocal opponent of Linnaeus in the 18th century, had distinguished 'natural' species (which are capable of fertile interbreeding) from 'artificial' species (as identified by Linnaeus). Darwin combines these two traditions so that the reality of 'species' is questioned and the idea of 'fixed' species is undermined; the distinction between 'species' and 'variety' is blurred. Darwin also affirms that species and varieties are defined by common descent. Varieties are 'incipient species'. To a knowledgeable contemporaneous reader, this chapter must have been disconcerting.

Chapter 3 generalises the Malthusian principle and broadens the original concept of competition for limiting resources in a novel way. Populations are controlled not only by the traditional restrictions on food and space but through *networks* of relationships among species: predator and prey, parasite and host and food-webs. Darwin's proposed mechanism is thus 'ecological'; it is not simply a matter of 'competition'.[14]

---

[13] The idea that certain features of a species are 'essential' is Aristotelian. Linnaeus may have superseded Aristotle's classification of animals, but not his way of thinking!

[14] So much for the suggestion that the *Origin* was simply an extrapolation to the natural world of the political economy of industrialised Victorian Britain and its spirit of competition. To update the wording (though not the substance) of Darwin's argument: the most successful variants of a species are those that maximise energy flow through an ecosystem. This 'thermodynamic' account of evolutionary mechanisms, emphasising 'cooperation' among populations within the ecosystem, has been developed particularly by the Japanese biologist Kinji Imanishi (1902–1992), whose cultural background is vastly different from Darwin's.

The natural selection model, with its enormous scope, is developed in Chapter 4. In the early editions of the *Origin* the status of natural selection is ambiguous. It may or may not be 'causal', or it may – in Aristotelian terms – be either efficient or final cause; so the explanation is still implicitly teleological. After the third edition, 'Nature' becomes no more than a literary metaphor and the implicit teleology disappears. Darwin's theory has at last become wholly *mechanistic*.[15]

The palaeontological evidence is not discussed in detail until Chapter 10, when Darwin, unlike Lyell, opts for a 'progressive' interpretation of the fossil record. The intervening chapters deal with the anticipated objections to the natural selection model: for instance, the existence of highly complex organs such as the mammalian eye, the evolution of animal instincts and of the 'social structures' of colonial insects. Various other objections are discussed in the later editions of the book.

A wide variety of issues in taxonomy, comparative anatomy, paleontology, biogeography and embryology are drawn together in Chapters 11–13 under the simple principles worked out in Chapters 1–4.

Application of the Natural Selection Model to Human Evolution

Darwin hardly mentioned human evolution in *The Origin of Species*; the speculations on this subject in *Vestiges* had unleashed a storm of criticism. But the topic could not be evaded. Archaeological discoveries of stone tools during the 1840s and 1850s led Lyell[16] and others to conclude that humans had existed for thousands of years prehistorically. However, there was no fossil evidence of the ancestors of modern humans until Java Man was discovered in the 1890s. Therefore, the debate that immediately followed the publication of *The Origin of Species* centred on the similarities and differences between humans and modern apes. Richard Owen vigorously defended the classification of humans as a completely separate order, as in the writings of Linnaeus and Cuvier. In contrast, Huxley sought to demonstrate a close anatomical relationship between humans and apes.[17] Lyell and Wallace agreed that humans shared a common ancestor with apes but questioned whether a Darwinian mechanism could account for the distinctive features of the human mind.

In 1871 Darwin published his views on human evolution in *The Descent of Man, and Selection in Relation to Sex*. He argued that the human mind did not differ qualitatively from the ape mind. Thus, morality was treated as an abstraction of instincts that were beneficial to all social animals. Darwin proposed that all the differences between humans and apes could be explained by a combination of the selective pressures resulting from our ancestors moving from the trees to the plains, and sexual selection. The debate about human uniqueness continues today.

---

[15] In the fifth edition (1869), Spencer's phrase 'survival of the fittest' is adopted as a synonym for 'natural selection', emphasising this metaphysical shift.

[16] Geological Evidences of the Antiquity of Man. John Murray, London (1863).

[17] Irvine W (1955) Apes, Angels and Victorians: the Story of Darwin, Huxley and Evolution. McGraw Hill, Columbus, Ohio.

The Philosophical Context

Darwin made no explicit reference to philosophies of science in his major works, but the central argument of the *Origin* points to the influence of Whewell. The natural selection model is established as a 'consilience of inductions'. However, that view may be simplistic. Whewell's ideas were essentially Kantian, and while Darwin's theory was generally well received in Germany, the most committed German Kantians, the mechanistic materialists, were sceptical of it. They perceived a taint of *Naturphilosophie* in Darwinism. Also, a major trend in the philosophy of knowledge in 19th century Britain was *positivism*. Darwin could hardly avoid its influence.

Positivism

Positivism is an empiricist philosophy, but it was not connected with Kant or the mechanistic reaction against *Naturphilosophie*. It was adopted initially by Auguste Comte (1798–1857) but was developed as a philosophy of science by John Stuart Mill (1806–1873).[18] It holds that the goal of knowledge is simply to *describe* observed phenomena, not to question their existence. All valid knowledge consists of descriptions of the evidence of the senses. Other areas of human knowledge and experience are subordinate to (and ultimately reducible to) science. Knowledge grows only by the addition of new or improved data. Progress is thus inherent in science; it is also inherent in society. This attempt to restore Enlightenment optimism was well received in Victorian Britain.

Mill did not share Hume's or Berkeley's scepticism. Things-in-themselves are real; when they are not actually being perceived by an observer they remain 'permanent possibilities of sensation'. Inductive reasoning is justified by the *principle of the uniformity of nature*, which is itself established inductively. Logical, deductive reasoning (from the general to the particular) is parasitic because the 'general' can only be established by induction from observation. Mill argued that Kant's 'innate structure of the mind', including the truths of mathematics and logic, had its origins in early learning.

---

[18] For Comte, positivism was the application of scientific methods to what we would now call 'sociology'. Comte's aim was to establish ways of controlling and manipulating society, just as Baconian natural philosophy was intended to establish ways of controlling and manipulating Nature. This precocious version of extreme socialism (or fascism?) was probably fostered by Comte's mentor, Saint-Simon (1760–1825). Mill dissociated himself from any such ambition. In his *On Liberty*, he wrote: '*M. Comte, in particular, whose social system, as unfolded in his* Système de Politique Positive, *aims at establishing (though by moral more than by legal appliances) a despotism of society over the individual, surpassing anything contemplated in the political ideal of the most rigid disciplinarian among the ancient philosophers...*' Mill's positivism was libertarian, not authoritarian.

It is easy to debunk positivism. For example, if every valid aspect of human knowledge is reducible to science, then either religion and ethics are reducible to science, or no ethical or religious proposition can be valid. If our senses are the only valid sources of belief, then the Earth is stationary and the sun orbits it. Also, human knowledge is not 'atomised', it is highly integrated. It grows according to rules of consistency, metaphorical extension, imaginative leaps and public debate, not just by adding new bricks to an existing wall. And even if the proposition 'Progress is inherent in society' were reducible to the evidence of the senses (it is not, so if positivism is correct, the proposition is not valid), it would be empirically false. Most societies (those of native Australians, for example) remain or remained static for uncounted generations.[19]

For these and other reasons, philosophers of science have long since rejected positivism. In mid-19th century Britain, however, it was tacitly or explicitly accepted by a fair consensus of scientists. It also had adherents in other countries, such as Helmholtz in Germany (see Chapter 9). Thus, although there are clear signs of Whewell's influence in parts of the *Origin* (notably in Chapter 4), there is a strong thread of positivism (for example in Chapters 1 and 2). Moreover, most proponents of Darwinism in later 19th century Britain were positivists.

The Fate of Positivism

Positivism attained an extreme form in the writings of the physicist Ernst Mach (1838–1916). According to Mach, the raw information acquired by the senses constitutes the *whole* of knowledge. Nothing is worthy of belief about the world unless it is justified by direct sensory evidence. During the 1920s, Mach's extreme positivism ('phenomenalism') became the basis for *logical positivism*, a briefly influential philosophy of knowledge.

The empiricist Rudolf Carnap (1891–1970) was for a time associated with the logical positivists. Carnap attempted to solve Hume's 'problem of induction' by formalising Mill's philosophy in a quasi-mathematical or statistical way. He argued that relative degrees of certainty can be attributed to different aspects of our knowledge according to the weight of sensory evidence supporting them. Carnap has not proved influential.[20]

Karl Popper identified fatal flaws in logical positivism, and indeed in the whole positivist tradition. Popper evaded Hume's 'problem of induction' by inverting it, observing that science progresses by proposing imaginative hypotheses and testing

---

[19] Some positivists, including Darwin's friend Herbert Spencer, believed that progress was inherent not only in nature and society but in the universe as a whole. We would now dismiss that as nonsense, but as we have seen in this chapter, it was crucial in the development of evolutionary theory in the mid-19th century.

[20] Carnap R (1936) Testability and meaning. *Philos Sci* 3:419–471. Creed I (1940) The justification of the habit of induction. *J Philos* 37:85–97.

them critically against observational or experimental evidence. No amount of evidence can *prove* a hypothesis, but one good piece of contrary evidence can *disprove* ('refute' or 'falsify') it. What we believe 'true' in science consists of hypotheses that have survived critical attempts at refutation. Many leading scientists approve of Popper's account. The best mindset for a scientist is Popperian; the aim is to test ideas critically, not to try to prove that they are right.

Acceptance of Evolutionary Theory and Natural Selection

Thomas Henry Huxley[21] (1825–1895) wrote in his essay on the reception of the *Origin*:

> The suggestion that new species may result from the selective action of external conditions upon the variations from their specific type which individuals present, and which we call spontaneous because we are ignorant of their causation, is as wholly unknown to the historian of scientific ideas as it was to biological specialists before 1858. But that suggestion is the central idea of the Origin of Species, and contains the quintessence of Darwinism.

In his campaign for public and scientific acceptance of Darwin's theory, Huxley relied heavily on newly-emerging palaeontological evidence, including the discovery of *Archaeopteryx*, consistent with the view that birds had evolved from reptiles; a line of reasoning on which Darwin himself had placed less emphasis.

   Within a few years of the *Origin*'s first publication, most scientists had, like Huxley, accepted the reality of evolution. However, the natural selection model was more controversial. Apart from orthogenesis (see earlier footnote), which retained a following for many years particularly among palaeontologists, significant numbers of people opted for a belief in theistic evolution or for some version of Lamarckism. These alternatives to natural selection could not solve the problem of purpose because they were all ineluctably teleological.[22] As we shall see in the next chapter, the pioneers of genetics opted for yet another alternative, *saltationism*, which was just as mechanistic as natural selection and led to a fierce debate about the mechanism of evolution.

---

[21] Huxley's advocacy of evolution and his drive to displace Paley's natural theology became a cornerstone of his endeavour to reform and professionalise science. Similarly, in Germany, Ernst Haeckel used Darwin's theory to challenge metaphysical idealism and *Naturphilosophie*. By the early 1870s, evolution had become the mainstream scientific explanation for the origin of species in the English-speaking world and in Germany. As we have seen (Chapter 10), it was less readily accepted in France and other countries.

[22] Teleological accounts of evolution have proved remarkably resilient – and scientifically useless. Pierre Teilhard de Chardin (1881–1955) proposed that the universe had developed gradually from subatomic particles to human society; an echo of Chambers and Spencer. His ideas were the indirect ancestor of the modern Gaia theory, which we discussed briefly in *About Life*. Henri Bergson (1859–1941) also proposed a 'progressive' version of evolution. His *Creative Evolution* (1907) centred on the concept of *élan vital* – vital force – which is strikingly similar to Schopenhauer's concept of the Will and to the ideas of *Naturphilosophie*. Several other ideas in Bergson's works are reminiscent of Schopenhauer.

The Age of the Earth

William Thomson, Lord Kelvin (1824–1907), a pioneer of classical thermodynamics and one of the most eminent scientists of the age, argued that the Earth was not old enough for the long-duration processes implicit in uniformitarian geology and the natural selection model of evolution.[23] For natural selection to account for the plethora of known species, past and present, the Earth had to be at least hundreds of millions of years old. Kelvin proved from the planet's mass and temperature that the upper limit of its age could hardly exceed ten million years, unless, as he observed in a careful caveat, there was some other as yet unknown source of heat. His student Henry Fleeming Jenkin (1833–1885) discounted this 'additional source of heat'. Jenkin also reasoned that small variations among individuals could not lead to species divergence, as Darwin claimed, because such variations would cancel each other out by interbreeding.[24]

Darwin was profoundly troubled by the Kelvin-Jenkin argument and never found an adequate answer. Kelvin's critique was the main reason for widespread scepticism about the role of natural selection in evolution during the late 19th century. Not until the end of that century, well after Darwin's death, was radioactivity discovered: Kelvin's alternative source of heat. This discovery was instrumental in establishing the now-accepted age of the Earth: some 4,500 million years, plenty of time for evolution by natural selection to have produced the observed abundance of extant and extinct species.

The Kelvin-Jenkin argument was probably motivated, at least in part, by commitment to orthodox Christian doctrine. In 1871, Kelvin wrote[25]:

> Reaction against the frivolities of teleology, such as are to be found, not rarely, in the notes of the learned commentators on Paley's 'Natural Theology,' has I believe had a temporary effect in turning attention from the solid and irrefragable argument so well put forward in that excellent old book. But overpoweringly strong proofs of intelligent and benevolent design lie all around us; and if ever perplexities, whether metaphysical or scientific, turn us away from them for a time, they come back upon us with irresistible force, showing to us through Nature the influence of a free will, and teaching us that all living things depend on one ever-acting Creator and Ruler.

This great physicist plainly believed that biology could not, and should not, be a science. He affirmed a perspective on the study of life that had been eliminated from physics two centuries earlier.

---

[23] On the secular cooling of the earth *Trans Roy Soc* Edinburgh 23:167–169 (1864); The doctrine of uniformity in geology briefly refuted. *Proc Roy Soc* Edinburgh 5:512–513 (1866); *On Geological Time*, address to the Geological Society of Glasgow (1868); On geological dynamics. *Trans Geol Soc* Glasgow (1869).

[24] Bulmer M (2004) Did Jenkin's swamping argument invalidate Darwin's theory of natural selection? *Br J Hist Sci* 37:281–297.

[25] *On the origin of life*. Report of the 41st Meeting of the British Association for the Advancement of Science; held at Edinburgh in August 1871, pp. 84–105.

The Nature of Heredity

Maupertuis and Buffon had established that inheritance involved the transmission of a material substance from parent to offspring (Chapter 10), just as Darwin's theory required. But Darwin's theory also required that offspring resemble their parents generally, but not exactly: the transmission process must be able to produce *variation*. What was the 'material substance' involved, how was it transmitted from parent to offspring, and how could it change so as to allow evolution by natural selection? Many biologists, including Darwin himself, expended much time and effort on these questions, but prior to the advances in cell biology that followed the innovations in microscopy in the 1870s and '80s, no satisfactory answers were obtained. Even after chromosomes were discovered, the nature of heredity remained a matter of intense debate. Only after that debate was settled could the theory of evolution take a further step towards maturity.

Chapter 13
The Great Heredity Debate

Darwin's Account of Heredity

Why do individuals within a species differ? The answer was crucial not only for the natural selection model but for any theory of evolution. According to the consensus 'inheritance of acquired characteristics' view, to which Darwin initially subscribed, variation is induced by the environment; a *positive* effect, improving adaptation. But natural selection assumes that the environment exerts a *negative* effect, eliminating the ill-adapted (see Fig. 12.1).

There was another crucial question: how are differences among individuals transmitted to their offspring? Darwin initially presumed some sort of 'blending inheritance' such that the characteristics of the offspring were 'averages' of those of the parents. Jenkin launched a cogent attack on this idea, and Darwin's own extensive studies during the 1860s led him to modify it. Among other things, he recognised that:

- There are occasional 'sports' in which bizarre deformities appear.
- Injuries are not inherited.
- Some characteristics are inherited by male but not female offspring.
- 'Throwbacks' occur, so that the offspring resemble, for example, grandparents rather than parents.

This motley array of observations seemed hard to explain. Darwin wrestled for years with the nature of heredity without reaching a satisfactory conclusion. In 1868, in his two-volume *Variation of Plants and Animals Under Domestication*, he adopted as a 'provisional hypothesis' the notion of *pangenesis*, which dates back to Hippocrates and is implicit in Buffon. It was the most effective reply he could find to Jenkin. According to Darwin's version of pangenesis, invisible material particles called 'gemmules' exist within cells and can be modified by environmental influences. They are shed continually into the blood stream and assembled by *'mutual affinity into buds or into the sexual elements'*, from which they are transmitted to offspring.

Darwin thought that pangenesis could explain (a) heritable variation, (b) the mechanism of Lamarckian use-disuse inheritance and (c) the inheritance of acquired characteristics. But few were persuaded. Wallace rejected the hypothesis and advised Darwin to do likewise.

The Biometric School

Nevertheless, Darwin's view of heredity (pangenesis and inheritance of acquired characteristics) had important adherents. It was developed by the 'biometric' school, initially associated with Darwin's cousin, Francis Galton (1822–1911), and subsequently with Karl Pearson (1857–1936) and Walter Weldon (1860–1906).

Galton conducted rabbit-breeding experiments in an attempt to support pangenesis. The results were unsatisfactory but led him to investigate inheritance statistically. His endeavours culminated in a sophisticated mathematical theory of inheritance, first outlined in 1889 and further elaborated in 1897,[1] when a 'statistical law of heredity' was proposed. With this mathematical law, both 'throwbacks' and persistent patterns of inheritance could be explained by calculating different 'strengths of ancestry' (the relative contributions of more or less distant ancestors to the characteristics of the individual). Although this achievement is now largely disregarded, Galton's work produced principles and methods that are important in modern statistics.

Galton's mathematical analysis of variation and the transmission of inheritance inspired Weldon, who was professor of zoology at University College London and later at Oxford. Weldon extended Galton's methods to natural populations of organisms in the wild. Initially, he agreed with Galton that the effects of natural selection were generally insignificant, but friendship and collaboration with Pearson changed his views. Pearson also appreciated Galton's statistical methods. Together with Weldon, he developed a mathematical analysis of variation in natural populations.[2]

Pearson was committed to positivism in the extreme form proposed by Mach (Chapter 12). This led him to reject any search for 'hidden causes' and to dismiss theoretical entities. Therefore, he could not countenance Darwin's 'gemmules' or pangenesis. However, he was willing to treat natural selection as a hypothesis worthy of testing. He and Weldon elaborated Galton's mathematical methods and applied them to characteristics of organisms that they could *measure*, searching for slight variations upon which selection might operate. Their analysis seemed to show that populations would gradually diverge under identifiable 'selection pressures', as Darwin had predicted. Being positivists, they made no assumptions about hidden underlying causes, so they inferred that long-term evolutionary change could be brought about by selection acting on slight individual differences.

---

[1] Natural Inheritance. Macmillan, London (1889); The Average Contribution of Each Several Ancestor to the Total Heritage of the Offspring. Harrison and Sons, London (1897).
[2] Norton BJ (1973) The biometric defense of Darwinism. *J Hist Biol* 6:283–316.

The biometricians emphasised that Darwinian natural selection operated on *continuous* variations of characteristics. Weldon showed that in real-world populations, selection pressure from the environment could indeed shift the range of variation. Thus, by taking a positivist stance and deploying sophisticated mathematics, the biometric school established strong support for the natural selection model of evolution.[3]

The Weismann Barrier

Weismann pondered pangenesis, and the inheritance of acquired characteristics, for many years. His early (1870s) acceptance of these proposals gradually (1890s) gave way to rejection. He retained his belief in Darwinian evolution by natural selection but denied the mechanisms of inheritance that Darwin had imagined. He proposed a *one-way* relationship between the hereditary material contained in the gametes (the *germ-plasm*) and the rest of the body (the *somatoplasm*). He recognised that the germ-plasm is continuous from generation to generation. In each generation it is responsible for forming the somatoplasm. Crucially, however, the somatoplasm does not influence the germ-plasm.[4] Heritable variations arise only when the germ-plasm is changed. The *unidirectionality* of this relationship between gametes and body is known as the 'Weismann barrier'.

Weismann's work was not fully appreciated until the 1930s and it has often been misinterpreted. In retrospect, it is tempting to equate the Weismann barrier with Mendelian genetics on the one hand and the central dogma of molecular biology on the other. In fact, Weismann was sceptical of Mendel's work, which was rediscovered and popularised in the early 20th century (see below). As for the central dogma of molecular biology, Crick told us that the genetic information in DNA is transcribed to messenger RNA, which is translated to a protein, but the information in a protein cannot be 'reverse-translated' into RNA and/or DNA. Although this dogma seems superficially similar to the Weismann barrier, it concerns the *expression* of genetic information, not its inheritance. In any case, we may cavil at the implicit equating of Weismann's 'somatoplasm' with 'proteins'.

Weismann's proposal was not strongly supported by experiment so it remained open to doubt. As Henri Bergson observed in 1907:

> ... if, perchance, experiment should show that acquired characteristics are transmissible, it would prove thereby that the germ-plasm is not so independent of the somatic envelope as has been contended... experience alone must settle the matter.

---

[3] Norton BJ (1973) ibid.

[4] The Germ-Plasm: a Theory of Heredity. Schribner, New York (1893). Weismann did not deny that the environment can alter heredity, but it can do so only by affecting the germ-plasm directly, not via the somatoplasm.

Mendel

Johann (Gregor) Mendel (1822–1884) showed academic ability in his childhood, but his peasant farmer family could not afford higher education for him. He therefore entered the Augustinian monastery at Brünn (Brno) to continue his studies. He had a flare for teaching as well as scientific investigation. In the early 1850s he studied mathematics, physics, botany and zoology at the University of Vienna, so he was thoroughly informed about the major scientific issues of the day. He never obtained a full teaching certificate, but that did not diminish his enthusiasm for study, particularly the study of plant hybrids. He believed that hybridisation was a key mechanism of evolution, a belief perhaps traceable to Linnaeus (Chapter 12).

Mendel was fascinated by the widespread belief that the environment induces heritable changes in organisms. Finding a variant form of a plant, he transplanted it to a new site, next to a normal ('wild type') plant of the same species. He cross-pollinated the two. The progeny had characteristics of both parents, no matter whether they were planted in the old site or the new one. Therefore, the environment does **not** induce heritable changes in *traits*. Also, it seemed that the traits of the offspring were determined by both parents, not just by the male as many people believed. These and similar observations inspired the experiments on pea plants that Mendel surveyed in his 1866 paper, which was later to become famous.

Plant hybridisation was a major topic of investigation in the mid-19th century and had interested Darwin. Mendel's copy of the German translation of the *Origin*, published in 1860, clearly inspired him. In one of his many marginal notes he commented that Darwin's notion of 'gemmules' had not received adequate attention. Large parts of his 1866 paper cite the same authorities that Darwin cites in Chapter 9 of the *Origin*; the same issues concerned him. For instance, Darwin wrote:

> Gärtner expressly states that hybrids from long cultivated plants are more subject to reversion than hybrids from species in their natural state; and this probably explains the singular difference in the results arrived at by different observers: thus Max Wichura doubts whether hybrids ever revert to their parent-forms, and he experimented on uncultivated species of willows... Gärtner further states that when any two species, although most closely allied to each other, are crossed with a third species, the hybrids are widely different from each other...

The second paragraph of Mendel's 1866 paper reads:

> Gärtner, especially in his work Die Bastarderzeugung im Pflanzenreiche, has recorded very valuable observations; and quite recently Wichura published the results of some profound investigations into the hybrids of the Willow. That, so far, no generally applicable law governing the formation and development of hybrids has been successfully formulated can hardly be wondered at by anyone who is acquainted with the extent of the task, and can appreciate the difficulties with which experiments of this class have to contend.

The penultimate paragraph of his paper shows how much significance he attached to the earlier studies:

> Gärtner, by the results of these transformation experiments, was led to oppose the opinion of those naturalists who dispute the stability of plant species and believe in a continuous evolution of vegetation. He perceives in the complete transformation of one species into

another an indubitable proof that species are fixed with limits beyond which they cannot change. Although this opinion cannot be unconditionally accepted we find on the other hand in Gärtner's experiments a noteworthy confirmation of that supposition regarding variability of cultivated plants which has already been expressed.

That 'supposition regarding variability of cultivated plants' refers to the results of Mendel's own experiments. The 'complete transformation of one species into another' was to become the central tenet of the saltationist or mutationist school (see below).

Over a period of eight years, Mendel conducted a *quantitative* study of seven clearly identifiable pairs of traits in some 29,000 pea plants in the monastery garden. He described his results in a lecture entitled 'Experiments in plant hybridisation', read to two meetings of the Natural History Society of Brno in 1865. This lecture was the basis of the 1866 paper published in the *Proceedings of the Natural History Society of Brno*.

Mendel found that traits do not blend or mix but remain constant over generations of crossing and hybridisation. They also remain separate. However, one of a pair of traits[5] may be 'dominant' over the other. For example, when pure-bred *tall* pea plants were crossed with pure-bred *short* ones, all the offspring were tall. When these tall hybrids were crossed with each other, three-quarters of the offspring were tall and a quarter of them were short. 'Tall' is the dominant trait; 'short' is recessive.[6] The first generation hybrids 'contain' equal amounts of the tall and short traits. When they form gametes (pollen and ova), each gamete has an equal chance of 'containing' either the tall or the short trait. At fertilisation, they combine randomly. Only if a short-trait pollen fertilises a short-trait ovum is the progeny plant short. All other combinations lead to tall progeny (Fig. 13.1).

Karl Wilhelm von Nägeli (1817–1891), who had been directed to the study of plant structure and hybridisation by Schleiden, was the leading authority in the field. He was not impressed by Mendel's results, which he thought unlikely to apply to other species. Indeed, Mendel's experiments on other plants gave much less clear-cut data.[7] His research was cut short when he became Abbot of the monastery in 1868; his new administrative duties left him with little time for research. By the time the significance of his findings was recognised, he had been dead for 15 years.

---

[5] Throughout his paper, Mendel referred to traits (*die Merkmale*) or characters (*die Charaktere*) but not to hypothetical underlying 'causes' of the traits/characters. He made occasional use of other words that have led to confusion on this point: *die Faktoren, die Elemente* and *die Anlage*. The first two of these ('factors' and 'elements') are most plausibly interpreted as the *components* of traits. The third is notoriously difficult to translate but is probably best understood as 'tendency' or 'predisposition'. The crucial point is that Mendel said nothing about 'genes' or anything that can legitimately be interpreted as 'genes'.

[6] The terms 'dominant' and 'recessive' are Mendel's. They refer to *traits* or *characteristics*. Many modern commentators write about 'dominant and recessive *alleles* (forms of genes)' – a nonsensical notion and a caricature of Mendel. See Porteous JW (2004). We still fail to account for Mendel's observations. *Theor Biol Med Mod* 1:4.

[7] Ever since Mendel's paper was 'rediscovered', comments have been made about the remarkable accuracy of his data. Statistically, it is highly improbable that such near-perfect 3:1 ratios could

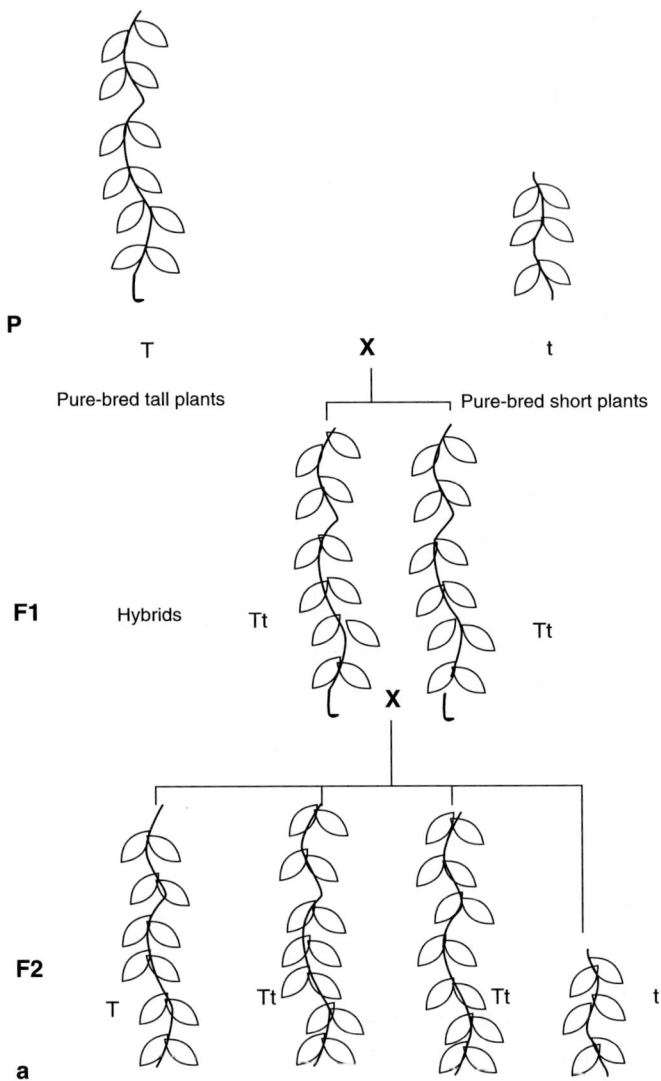

Fig. 13.1 Schematic illustration of Mendel's hybridisation experiments. (a) When pure-bred tall pea plants (trait T) were crossed with pure-bred short plants (trait t), all the progeny were tall. Mendel reasoned that these hybrids must contain *both* traits, so they were marked Tt, but the tall trait is *dominant* and the short trait is *recessive*. (The symbols P = parental generation and F1 = "first filial" generation are due to Bateson, not Mendel.) When the F1 hybrids were crossed with each other, three-quarters of the progeny ("second filial" generation, F2) were tall (T or Tt) and one-quarter were short (t).

have been obtained from unbiased experiments. The implication is that the good Abbot cheated. People who draw this inference should remember that the 1866 paper was not a scientific paper as we understand it today, but the transcript of a *lecture*. When we deliver lectures, we aim for clarity, not for an unbiased and detailed survey of all our data (in this case, on 29,000 plants). We select the best results in order to make our conclusions clear to a largely non-specialist audience.

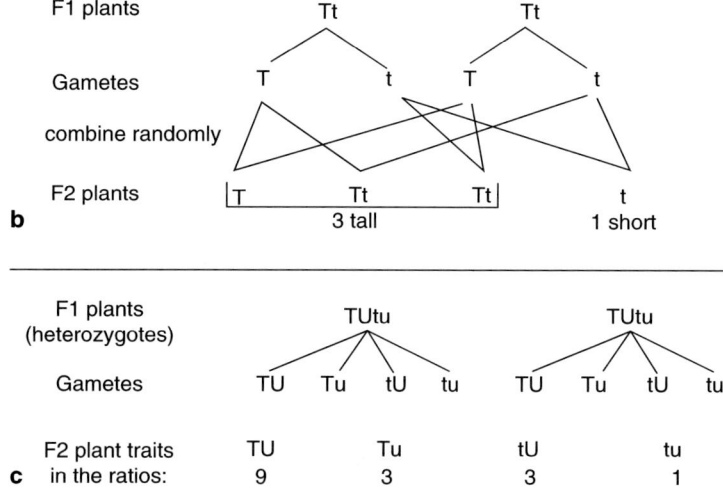

b

c

Fig. 13.1 (continued) (b) Mendel explained these findings, and similar findings for other traits, on the basis of random assortment of gametes. Each gamete can 'contain' only one trait, in this case T or t. (c) Two different traits are sorted independently when F1 hybrids from pure-bred P strains are crossed. Here the traits are symbolised by T (dominant), t (recessive); and U (dominant), u (recessive). Each gamete contains either T or t *and* either U or u. Random recombination of the gametes gives progeny of which 9/16 have both dominant traits (T and U), 6/16 have one dominant and one recessive (T and u or U and t), and 1/16 have both recessives (t and u). To avoid confusion, the sixteen possible combinations are not shown in detail

The Mutationist (Saltationist) School

In 1894, the zoologist William Bateson (1861–1926), who coined the word 'genetics', published a large study of the empirical evidence favouring either continuous or discontinuous variation in natural populations.[8] He concluded that natural species were separated by fundamental *discontinuities*. This was contrary to the conclusions of Pearson and Weldon and also, it seemed, to Darwinism. Ironically, Bateson was a former student of Weldon. He wrote:

> the Discontinuity of which Species is an expression has its origin not in the environment, nor in any phenomenon of Adaptation, but in the intrinsic nature of organisms themselves, manifested in the original Discontinuity of Variation.

Like the biometricians, he offered no causal explanation of individual differences, but his detailed evidence demanded further investigation of the continuity/discontinuity issue.

In Amsterdam at this time, Hugo de Vries (1848–1935) was performing a series of botanical experiments, mainly on wild varieties of the evening primrose, and

---

[8] Materials for the Study of Variation: Treated with Special Regard to Discontinuity in the Origin of Species. Macmillan, London.

formulating empirical laws of discontinuous inheritance. De Vries revised Darwin's idea of gemmules; he called them 'pangens' and asserted that they were not transported around the body as Darwin had said but remained within the cells. Each 'unit character' in an organism was associated with a single pangen. Thus, an organism is a mosaic of characters, each determined by a pangen. De Vries concluded that evolution took place not gradually, as Lamarck and Darwin had held, but in 'jumps' (*saltations*) caused by qualitative changes (*progressive mutations*) in pangens and therefore in traits. (De Vries also recognised 'retrogressive' and 'degressive' mutations, which occurred in hybridisation and could not be 'fixed' by natural selection.) The German botanist Karl Correns (1864–1933), a student of Nägeli, made similar observations in his study of hawkweed hybrids.

Shortly before de Vries published his work (1900), his attention was drawn to Mendel's 34-year-old paper, which seemed consistent with his new laws. Correns also knew about Mendel's paper because of the long and unproductive correspondence between his mentor, Nägeli, and Mendel during the 1860s, and he too cited it.[9] Like Mendel, he recognised that in many (though not all) pairs of traits, one trait is dominant and the other recessive. Unfortunately, Correns believed that the recessive trait is 'suppressed' by the dominant trait in hybrids, an unfounded presumption.

Bateson saw that his own experimental findings were consistent with those of Correns, de Vries and Mendel,[10] so he began to promote Mendel's work as a general explanation not merely of hybridisation but of inheritance and thus of evolution. As Bateson and the saltationists saw it, Darwin was wrong: evolution is driven by stepwise changes in traits underpinned by pangens, not by the gradual process of natural selection.

The Great Heredity Debate

A bitter dispute ensued between biometricians and the mutationists (saltationists). The simple algebra used by Bateson and de Vries was no match for the sophisticated mathematical and statistical techniques developed by Pearson and his students. Indeed, Bateson and his colleagues found it very hard to deal with those complex mathematical arguments. There was another important difference: Bateson, de Vries and Correns studied controlled populations under experimental conditions, while the biometricians studied unselected wild populations.

---

[9] Correns later discovered 'cytoplasmic inheritance' in plants: as we now know, chloroplast genes are inherited from the female parent and are independent of the nuclear genes. (Mitochondrial DNA, like chloroplast DNA, is also inherited from the female parent, so mitochondrial genes too give rise to 'cytoplasmic inheritance').

[10] The Austrian agronomist Erich von Tschermak (1871–1962) has also been credited with the 'simultaneous rediscovery of Mendel's laws' in 1900. Some historians have doubted whether Tschermak understood Mendel's work. He was certainly much less interested in the theoretical implications of Mendel's discoveries than de Vries, Correns or Bateson; instead, he saw their practical value and applied them to the cultivation of new crop strains.

Weldon and Pearson showed mathematically that Mendelian inheritance could be a special case of blending inheritance, as described by Pearson's modifications of Galton's law. Thus, the data supported Darwin's theory. The mutationists, in contrast, maintained that the variations studied by biometricians were too trivial to account for the evolution of new species. Only 'mutations', i.e. changes in de Vries's 'pangens', could introduce genuine novelty into breeding populations. If many pangens contributed to a measurable trait, then the continuous variation described by the biometricians might be observed.

There seemed to be sound arguments on both sides. Yet the new experimental approaches adopted by de Vries and Bateson won wide international allegiance very quickly. Why?

The general dislike of sophisticated mathematics among biologists may have contributed; the simpler the maths, the more likely your ideas are to attract support. Another factor was a discovery made independently by the mathematicians Godfrey Hardy (1847–1947) and Wilhelm Weinberg (1862–1937) in 1908: if a stable population undergoes random mating, then assuming Mendelian inheritance, the proportions of dominant and recessive traits will remain stable. This *Hardy-Weinberg Equilibrium* showed that Mendel's rules applied to wild populations as well as experimental ones.

But the main factor was the new cell biological work on chromosomes.

Chromosomes and Heredity

Weismann, Wilson and others had suggested that chromosomes were or contained the 'material substance of heredity' proposed by Maupertuis and Buffon in the 18th century. By 1900, many people believed that chromosomes were permanent structures and were transferred to the daughter cells at cell division (Chapter 10). Boveri, an early proponent of this view, showed in 1902 that sea urchin zygotes with abnormal numbers of chromosomes developed abnormally. This broadly supported the chromosomal hypothesis of inheritance. Stronger corroboration appeared in two papers[11] (1902 and 1903) by a student of Wilson, Walter Sutton (1877–1916).

Sutton studied dividing cells in grasshopper testes under the microscope and saw that the 46 chromosomes differed in size. Repeated observations showed that the 46 comprised 22 pairs of 'identical twins' (homologous pairs) and two that were unmatched, later identified as the sex chromosomes. This distribution of chromosome sizes and shapes remained the same at each *mitotic* cell division. But during *meiosis*, the special ('reductive') cell division process that produces spermatozoa, the chromosome number was halved.[12] Each gamete received precisely one of each chromosome pair, so it contained 23 non-identical chromosomes, not 46 (Fig. 13.2).

---

[11] (1902) On the morphology of the chromosome group in *Brachystola magna*. *Biol Bull* 4:24–39; (1903) The chromosomes in heredity. *Biol Bull* 4:231–251.

[12] Independent evidence was published in 1901 that chromosomes exist in 'homologous pairs', one derived from each parent. Montgomery TH (1901) A study of the chromosomes of the germ-cells of metazoans. *Trans Am Philos Soc* 20:154–236.

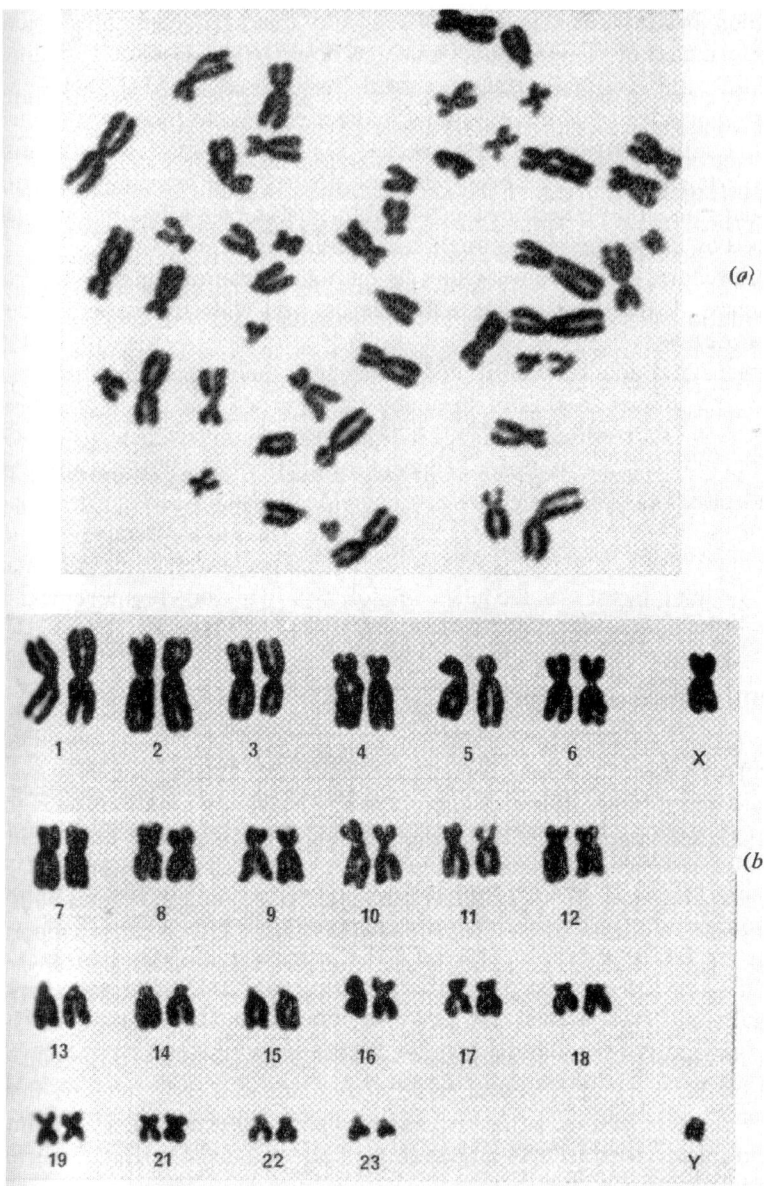

Fig. 13.2 Human chromosomes. The chromosomes of a normal human male. A dividing cell (at the metaphase stage of mitosis: see Chapter 10) is squashed between glass slides and the condensed chromosomes are spread out and then stained. This photomicrograph is magnified about 2,000 times. The upper picture (a) illustrates such a metaphase chromosome spread. In the lower picture (b), the chromosomes from (a) have been arranged in pairs, as Sutton did with the grasshopper chromosomes. In the case of the human male, there are 23 pairs of chromosomes, numbered in order of decreasing size; and a 24th 'unmatched' pair, the sex chromosomes, X and Y. In normal females, there are two X chromosomes and no Y. Reproduced with permission from Pearson Education from Fig. 1.5 (p. 17) in Haggis GH, Michie D, Muir AR, Roberts KB, Walker PMB (1964) Introduction to Molecular Biology. Longmans, London. Copyright Pearson Education

Sutton saw a striking parallel between these patterns of chromosome distribution during cell division and spermatogenesis and the Mendelian pattern of inheritance of traits. Thus, the hypothesis that chromosomes carry heritable material became more credible. However, if each chromosome represented one trait or character, then the number of traits of an organism must be equal to the number of chromosomes; but a grasshopper has more than 23 recognisable traits! Therefore, *several* traits must be represented on each chromosome. But if that is the case, then some traits cannot be independently assorted when gametes are formed. In short, Sutton's work supported the hypothesis that chromosomes carried the 'material substance' of heredity, but it also led to a new prediction: *only traits represented on* different *chromosomes could be independently assorted as described by Mendel.* As we observed in Chapter 1, a good scientific hypothesis leads deductively to clear, specific and *testable* predictions.

For a time, few were persuaded. Bateson and his colleagues, notably Edith Saunders (1865–1945), were among the sceptics. Bateson would not be converted chromosomal hypothesis of inheritance for another 20 years. Nevertheless, Sutton's work was a blow to the biometricians. Meanwhile, Bateson, Saunders and other investigators found broadly Mendelian inheritance patterns in many other organisms. Not all pairs of traits were clearly dominant/recessive, and there were other complications. However, during the first decade of the 20th century, 'Mendelian inheritance' was established as the norm, and saltation rather than natural selection was accepted as the mechanism of evolution. Bateson and Saunders captured this growing belief in a new terminology, which is more or less retained today[13]:

> We thus reach the conception of unit-characters existing in antagonistic pairs. Such characters we propose to call *allelomorphs*, and the zygote formed by the union of a pair of opposite allelomorphic gametes, we shall call a *heterozygote*. Similarly, the zygote formed by the union of gametes having similar allelomorphs, may be spoken of as a *homozygote*.

Sutton's Errors

In Mendel's symbol system, pure-bred lines with dominant and recessive traits (homozygotes) were written *A* and *a* respectively while hybrids (heterozygotes) were written *Aa*. Sutton revised this system and wrote *AA* and *aa* for the homozygotes, thus confusing the traits (which are single measurements) with the paired chromosomes allegedly responsible for them.[14] He explicitly said that the chromosomes or chromosome components were the 'determining factors' of the traits. This confusion of what we now call 'genotype' with 'phenotype' has persisted. Sutton's unwarranted alteration of Mendel's symbol system is now common in textbooks.

---

[13] Reports to the Evolution Committee of the Royal Society. Harrison, London (1902). Notice that Bateson and Saunders tacitly *identify* their 'allelomorphs', which we now call *alleles*, with the traits or characters, compounding Sutton's confusion between 'genotype' and 'phenotype'.

[14] He actually used italics for the chromosome components and plain text for the traits, but he did use the same letters. This led to the confusion that still persists.

The word 'gene' (derived from de Vries's 'pangen' and hence from Darwin's 'pangenesis'), and the clear distinction between *genotype* and *phenotype*, were introduced in 1909 by the Danish biologist Wilhelm Johannsen (1857–1927).[15] He studied traits in self-fertilising beans and found that continuous variation of the kind studied by the biometricians could be resolved into discontinuous traits showing Mendelian inheritance. The clarity of Johannsen's genotype-phenotype distinction has been obscured by Sutton's confusion of symbols, and by the widespread misapplication of the adjectives 'dominant' and 'recessive' to alleles (genotype) as well as traits (phenotype).

Although Bateson rejected the chromosome hypothesis, he subsequently[16] followed Sutton's lead in using double letters for traits in homozygotes:

> In cases where the pure dominants are recognisably distinct from the heterozygous dominants, it must naturally be supposed that two "doses" of the active factor are required, one from the paternal and another from the maternal side, in order to produce the full effect.

Bateson assumed that this 'active factor' was absent in homozygous recessives. It is regrettable that Johannsen's work was not more thoroughly read and understood.

Morgan and the Chromosomal Theory of Inheritance

Thomas Hunt Morgan (1866–1945) studied under T. H. Huxley and then under H. Newell Martin, from whom he acquired an understanding of experimental physiology and a mechanistic materialist outlook. This philosophical stance was no doubt challenged by his exchange of views with Driesch and his debate with Loeb about embryology, but it persisted. Morgan worked closely with Wilson and acquired skills in cell biology. He made significant contributions to embryology, and he wrote a definitive book on the regeneration of parts of organisms. Even without his work on genetics, therefore, he would have been a significant contributor to biology. Nevertheless, he is most famed for constructing 'classical genetics' during the second decade of the 20th century. His students Calvin Bridges (1889–1938), Herman Muller (1890–1967) and Alfred Sturtevant (1891–1970) continued to work with him and were major contributors to this achievement.

In keeping with his mechanistic materialism, Morgan was a hard-line experimentalist, impatient with theory and dismissive of speculation. Until 1909 he was contemptuous of Weismann, rejected Darwin and did not believe Sutton's chromosomal hypothesis. He was highly critical of the Mendelian explanations that were

---

[15] Elemente der exakten Erblichkeitlehre mit Grundsatzen der biologische Variationstatistik. Gustav Fischer, Jena.

[16] On Mendel's heredity of three characters allelomorphic to each other. *Proc Cambridge Philos Soc* 12:153–154.

being advanced for more and more observations. In a talk to the American Breeders' Association in 1909 he remarked:

> If one factor will not explain the facts, then two are invoked; if two prove insufficient, three will sometimes work out... the results are often so excellently "explained" because the explanation was invented to explain them.

He was persuaded by the experimental findings of de Vries and the saltationist view of evolution, and although he did not speculate about 'pangens', the idea of a *particulate* basis for heredity appealed to him. Mechanistic materialists were always drawn to the notion of 'atoms' and their motions and interactions (Chapter 9).

In 1908–1909, economic circumstances compelled Morgan to start experimenting with a cheap, easily-kept and rapidly-reproducing species, the fruit fly *Drosophila melanogaster*.[17] His initial aim was to show that new *species* could be created by mutation alone under laboratory conditions, as de Vries had said. But within a few years his opinions had been transformed. The fact that *Drosophila* only has four pairs of chromosomes (three autosomes and one sex chromosome) was to prove highly convenient, though the significance of this was not suspected at the outset of Morgan's studies.

Between 1909 and 1915, Morgan, Bridges, Muller and Sturtevant studied changes in more than 100 visible *Drosophila* traits. All these changes were attributable to mutations. Contrary to expectation, most mutations had relatively small effects on the phenotype: a change in eye colour, for instance. Therefore, they increased the variation in the fly population but did *not* give rise to new species by saltation. The debate about the mechanism of evolution had not been satisfactorily resolved after all.

The *Drosophila* mutations sometimes followed a Mendelian inheritance pattern, but sometimes they did not. As Bateson and Saunders had found, the difference between mutant and wild type traits was not always 'recessive' versus 'dominant'; there was often no clear 'dominance'. More interestingly, some results corroborated Sutton's prediction: several characters did *not* assort independently, but were *linked*. Further investigation showed that two characters were linked if and only if their determinants were carried on the same chromosome. Indeed, all the *Drosophila* traits examined by Morgan's team could be classified into four 'linkage groups' that appeared to correspond to the four chromosome pairs. In one of these four groups, the characters were sex-linked (i.e. characters altered by mutation were much more likely to be manifest in male than female offspring), just as colour-blindness is in humans. Indeed, the discovery that one particular mutant trait, white eye colour, is sex-linked was instrumental in convincing Morgan of the chromosomal hypothesis of inheritance. It later transpired that the relevant gene for eye colour is located on the X chromosome.

---

[17] This is the preferred modern name of a species that remains the favourite object of study among geneticists. Biologists in the early 20th century generally followed Lowe's nomenclature and called it *Drosophila ampelophila*, which is the name used in Morgan's publications. There are several other synonyms.

Contrary to the speculations of de Vries and Sutton, the relationship between trait and chromosomal determinant (gene) proved not to be one-to-one. Several genes could be involved in determining one character; on the other hand, one gene could contribute to several characters. Not surprisingly, Sutton had underestimated the subtlety and complexity of the genotype-phenotype relationship.

Sutton had also been unaware of a quirk of meiosis called *crossing-over*, which was provisionally identified in 1909 but would not be confirmed experimentally until 1931.[18] Interestingly, Morgan accepted crossing-over as early as 1911. Mechanistic materialist and hard-line experimentalist he might be, but he needed the crossing-over mechanism to explain some puzzling data; therefore, he believed it.[19] The 'puzzling data' concerned traits that seemed to be partly but not completely linked: that is, they sometimes sorted independently and sometimes did not. Crossing-over afforded a plausible mechanism for this phenomenon, but it remained hypothetical until Stern's cell-biological studies in 1930–1931 confirmed it. Nevertheless, Morgan was willing to assert in 1911 that genes have definite, fixed locations on chromosomes, and that they are arranged in linear order. His philosophical predisposition towards 'material particles' led him beyond the evidence, though we now accept his 1911 assertions as essentially correct (Fig. 13.3).

The crossing-over hypothesis led Sturtevant and, subsequently, Bridges to perform *chromosome mapping* studies, assigning individual genes to specific chromosomal locations. This great achievement represented the coming of age of 'classical genetics'. It was to prove a vital step in the further maturation of the theory of evolution.[20]

Philosophical Problems and an Alternative Viewpoint

History, notoriously, is written by the winners, and Morgan and his colleagues were assuredly winners. Even the sceptical Bateson was convinced of the chromosomal theory of heredity by the early 1920s, and within a decade a consensus of biologists had accepted classical genetics. Stern's definitive proof of crossing-over in 1931 clinched the matter. But this triumph, and its undoubted importance in the history

---

[18] Janssens FA (1909) La théorie de la chiasmatypie. Le Cellule 25:387–406. Stern C (1931) Zytologische-genetische Unterschungen als Beweise für die Morganische theorie des Faktorenaustauchs. Biologisches Zentralblatt 51:547–587.

[19] It was technically impossible to confirm crossing-over experimentally in 1911, but Morgan was not straitjacketed by his own dictum: '*It is the prerogative of science… to cherish those theories that can be given an experimental verification and to disregard the rest, not because they are wrong, but because they are useless*'.

[20] Although Morgan long remained sceptical of Darwinism. Classical genetics was fully described in Morgan, TH, Sturtevant AH, Muller HJ, Bridges CB (1915) The Mechanism of Mendelian Heredity. Holt, New York; and in Morgan TH (1919) The Physical Basis of Heredity. Lippincott, Philadelphia.

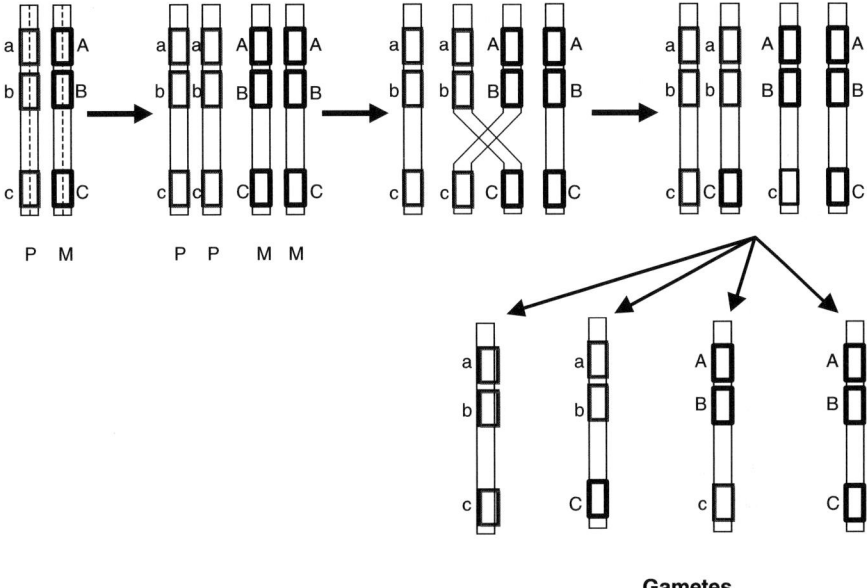

Gametes

Fig. 13.3 Crossing-over in a homologous chromosome pair during meiosis. Crossing-over occurs during the later stages, particularly the diplotene stage, of the complex 'prophase 1' of meiosis (see Fig. 10.4). By this stage, each pair of homologous chromosomes is aligned (P = paternal, M = maternal) and each chromosome has divided longitudinally (dotted lines) into two *chromatids*. Three pairs of alleles are shown: A, a; B, b; C, c. The paternal chromosome contains alleles a, b and c, while the maternal contains A, B and C. As prophase progresses, one paternal and one maternal chromatid cross over. As a result, four different combinations of the alleles of these three genes are found in the gametes: ABC, ABc, abC and abc. Genes A (or a) and B (or b) are close to each other on the chromosome so they are unlikely to be separated during crossing-over. Gene C (or c) is distant from the other two genes and is therefore much more likely to be separated from them during crossing-over. The diagram is adapted from an illustration in the 1915 book by Morgan and colleagues (Footnote 20)

of biology, should not blind us to certain philosophical shortcomings in Morgan's work, or to the fragility of the claims of classical genetics during the second decade of the 20th century.

Morgan's conception of genes and chromosomes was static, mechanical and morphological. Of course, that was entirely in line with the mechanistic view of physiology inherited from du Bois Reymond. In essence, however, the idea that genes are fixed particles in fixed chromosomal locations is *preformationist*. Since experimental embryology – to which Morgan himself had made significant contributions – had unequivocally confirmed the *epigenetic* view of development, that seems strangely inconsistent. Potentially, it created problems for the rapprochement between genetics and embryology.

In opposition to Morgan and his colleagues, the saltationist Richard Goldschmidt (1878–1958) proposed an epigenetic, dynamic and physiological view of genetics.

Goldschmidt did not challenge the experimental findings of Morgan's group but he doubted that crossing-over explained their data, so he was sceptical about chromosome mapping. Also, he believed that chromosomes disintegrate in the non-dividing nucleus. We dismiss these opinions today, but there was no *definitive* evidence for the durability of chromosomes or for fixed linear arrangements of genes around the time of the First World War.

For Goldschmidt, the chemical nature of the chromosome rather than its morphology determined heredity. The chromosome was a passive framework to which the hereditary determinants were attached. As embryos developed, different subsets of these determinants were released into the cytoplasm to alter the physiology of each type of cell. Goldschmidt suggested that the determinants, the genes, were enzymes. He wrote[21]:

> There must, therefore, be some kind of force that comes into play in the formation of the chromosome such that it always instructs each particulate hereditary factor to find its proper chromosome and its proper place again.

His aim was to show that *'the same forces that cause the individuality of the chromosomes can also explain crossing-over'*. These 'forces' were the basis of interactions among macromolecules, which pioneering biochemists were beginning to explore.

Morgan's colleagues devoted considerable energy to rebutting Goldschmidt's arguments, but not until the middle of the century was the consensus of biologists persuaded that genes were DNA, not enzymes.

There are two ironies in this historical vignette. First: neither Goldschmidt nor the classical geneticists fully appreciated the significance of Johannsen's genotype-phenotype distinction. Goldschmidt portrayed the genes as directly responsible for phenotypic characters, migrating from nucleus to cytoplasm and back again, changing the physiology of the cell as they moved. Morgan and his colleagues perpetuated the canard of describing *alleles* (different variants of a gene) as 'dominant' and 'recessive'. They 'reasoned' that if one of the alleles in a heterozygote is dominant and the other recessive, the character expressed in that heterozygote will be dominant. Since the designation of an allele as 'dominant' depended on the dominance of the trait that it determined, that was circular reasoning with a vengeance. And so it remains.

Secondly: the scientist who obtained the definitive proof of crossing-over, Stern, was a student of Goldschmidt.

---

[21] Goldschmidt R (1917) Crossing over ohne Chiasmatypie? *Genetics* 2:82–95.

Chapter 14
Evolutionary Theory Attains Maturity

Ronald Fisher (1890–1962) and the Foundation of Population Genetics

Early in the debate between the biometricians and the Mendelian saltationists, Pearson's former student George Yule (1871–1951) attempted to reconcile the two sides. He failed because the statistical methods available were inadequate. Two decades later, while Morgan's team was mapping the *Drosophila* chromosomes, Ronald Fisher began[1] to introduce new statistical techniques, many of which became standard. Fisher succeeded where Yule had failed: he assimilated Mendelian genetics into Darwin's natural selection model of evolution. These studies culminated in 1930.[2]

In his 1918 paper, Fisher critically analysed Pearson's claims. Contrary to Pearson, he showed that dominance is better explained by discrete Mendelian traits than by blending inheritance. By 1922 he had adopted Johanssen's term 'gene'. His critique of the biometric school made him appear anti-positivist; indeed, his philosophical position seemed close to Morgan's. He was never explicit about mechanistic materialism and he is remembered as a theorist not an experimentalist, but he used a deliberate analogy with physics to unify genetics with natural selection: '… *the whole investigation may be compared to the analytical treatment of the Theory of Gases*'. Like a physicist, he worked with idealised initial conditions, appealed to theoretical entities, and sought mathematical laws to encapsulate the phenomena of heredity and evolution.

By 'the theory of gases' Fisher meant *statistical mechanics*, a theory developed in the early 20th century. Statistical mechanics treats an isolated volume of a substance, usually a gas, as an ensemble of numerous particles (molecules), each with its own mass and velocity. The particles move *independently*, at random, colliding with each other and the walls of the vessel. Individual collisions can be described by Newtonian mechanics, but the behaviour of the whole ensemble can only be

---

[1] (1918) The correlation between relatives on the supposition of Mendelian inheritance. *Phil Trans Roy Soc Edinburgh* 52:399–433.

[2] The Genetical Theory of Natural Selection. Clarendon, Oxford.

described statistically. By analogy, Fisher constructed a 'statistical genetics,' applying statistical rules to combinations of alleles in a population. This model was fully developed in his 1930 book.

As Sutton had argued, and Morgan and colleagues had shown, genes that are linked on the same chromosome do not sort independently. Fisher's model takes account of that. It applies to a *population* of organisms of the same species, not to a mating pair and their offspring. The population is 'reproductively isolated', just as an ensemble of molecules in statistical mechanics is thermodynamically isolated. Any male in the population can mate with any female and vice versa. The frequencies of alleles within the population's *gene pool* can then be calculated and probabilities of survival at different life stages can be determined. Using this approach, Fisher turned key evolutionary concepts such as reproductive value and 'fitness' into quantitative, *measurable* terms.

In 1930 he stated his 'fundamental theorem of natural selection':...*the rate of increase in fitness of any organism at any time is equal to its genetic variance in fitness at that time*. He drew a parallel between this theorem and the second law of thermodynamics:

> ...both are properties of populations, or aggregates, true irrespective of the nature of the units which compose them; both are statistical laws; each requires the constant increase of a measurable quantity, in the one case the entropy of a physical system and in the other the fitness...of a biological population.

However, Fisher's fundamental theorem differs from the second law of thermodynamics in several ways:

- Species extinction has no analogue in physics.
- Fitness is different for every organism, whereas all molecules in an ensemble are considered identical.
- Fitness changes when the environment changes.
- The second law leads to increased disorder (higher entropy), while '*evolutionary changes are generally recognized as producing progressively higher organization in the organic world*'. As Schrödinger put it in his celebrated essay *What is Life?*: '*Life feeds on negative entropy*'.

For Darwin and the biometricians, evolution operated on the observable characters of *individual* organisms. Fisher recognised that although natural selection is a precise, highly non-random process, alleles are distributed through a *population* by random mating. He therefore modelled evolving populations as 'bags of atomic genes', which behave like ensembles of gas molecules in statistical mechanics. Thus, he reconciled the discontinuous character of Mendelian genetics with the gradual, continuous nature of evolution by natural selection, and he did so in a mathematically rigorous way. It was an outstanding and durable achievement.

He made other important contributions to biology. Following the chromosome mapping work of Morgan's group, he found statistical ways (the maximum likelihood method) of estimating degrees of genetic linkage. His 1950 paper on the estimation of gene frequencies represents the first application of computers in biology. He developed the Darwinian idea of sexual selection. Famously, he showed that the probability that a mutation can increase the fitness of an organism decreases in proportion to the magnitude of the mutation. He also proved that larger populations carry more variation than small ones and are therefore more likely to survive.

Fisher's Errors

Unfortunately, Fisher not only perpetuated the error (Chapter 13) of applying the descriptions 'dominant' and 'recessive' to alleles as well as traits, he added a bizarre twist to it. First, he made the unjustified assumption that a wild type trait is usually dominant. Therefore, if a (recessive) mutant arises, there must be mechanisms for bringing it back to the dominant wild type. He explained these imaginary mechanisms by proposing a separate set of 'modifier genes', which must be immune to mutation otherwise they could not work. (Also, they could only operate in heterozygotes.) Of course, no such implausible 'modifier genes' have ever been found – and how could they be 'immune' from mutation!

Why did so brilliant a contributor to biology (and statistics) propose such a strange idea? The answer may be related to Fisher's social and political views. He regarded less fortunate members of society as the victims of recessive alleles. He was a vigorous advocate of eugenics and believed that certain races of humans were biologically superior to others.[3] Such fanciful opinions were not exceptional among academics prior to the Second World War, but Fisher was an extreme proponent and remained so until late in his life.

Sewell Wright (1889–1988)

Sewell Wright was another highly gifted mathematician who invented several statistical techniques.[4] Like Fisher, he succeeded in unifying Mendelian genetics with natural selection, but he added the idea of *genetic drift*: random changes in gene frequencies can arise from random births, deaths and segregation of traits. He described the relationships among phenotypes in terms of *fitness surfaces* or *fitness landscapes*. These are 'three-dimensional maps' in which fitness is the height, and the horizontal axes are allele frequencies and average phenotypes. A population

---

[3] To the end of his life, Fisher was willing to defend unpopular and implausible claims. For example, he insisted on statistical grounds that smoking does not cause lung cancer (Smoking. The Cancer Controversy: Some Attempts to Assess the Evidence. Oliver & Boyd, Edinburgh). These aspects of his writings are overlooked by some commentators.

[4] These included the F-statistic, now a standard tool in population genetics, and an imaginative method known as *path analysis*.

occupies an isolated hill or peak on this map. Natural selection causes the population to climb a peak, while genetic drift may lead to wandering from one peak to another, passing through (maladaptive) 'lower ground' on the way.

Genetic drift is effective only on *small* populations. It allows genes to 'flow' to other populations, enabling adaptations to spread. Fisher rejected the idea, maintaining that variation could only be significant in *large* populations, and he and Wright engaged in a long and acrimonious dispute. For many years, genetic drift was regarded as unimportant in evolution, but opinions have changed recently.

The disagreement between Wright and Fisher had other dimensions. Apparently influenced by Goldschmidt (Chapter 13), Wright thought that genes may be enzymes. By 1917 he had modified this position and proposed that genes *control* enzymes[5]:

> ...it is clear that dominance has to do with the physiology of the organism and has nothing to do with the mechanism of transmission, i.e. with heredity in the narrow sense.

That was directly opposed to Fisher's view. Also, there seems to have been a philosophical difference between the two men. While Fisher's stance implied mechanistic materialism, Wright's was closer to positivism. He wrote:

> It is the task of science, as a collective human undertaking, to describe from the external side, such statistical regularity as there is in a world in which every event has a unique aspect, and to indicate where possible the limits of such description. It is not a part of its task to make imaginative interpretations of the internal aspect of reality.

J. B. S. Haldane (1892–1964)

Haldane, son of the 'holistic materialist' J. S. Haldane (Chapter 9), also studied genetics as a young man. He contributed significantly to human genetics and our understanding of enzyme function. Like Wright, he saw that genes and enzymes must be related; and like both Wright and Fisher, he found mathematical ways of reconciling Darwinism with Mendelism. His model showed that both the *direction* and the *rate* of gene frequency changes in populations can be determined.[6] He also considered the effects of immigration and emigration, i.e. he assumed populations that were not 'ideal' or 'isolated' in Fisher's sense. He was aware of the shortcomings of Fisher's explanation of dominance as early as 1930 and favoured Wright's more reasonable interpretation.

---

[5] Wright S (1934) Physiological and evolutionary theories of dominance. *Am Naturalist* 63:24–53. This view of dominance was far more biologically (and logically) plausible than Fisher's, though in Wright's day too little was known about enzymes and metabolism for the statement to be made more precisely.

[6] (1932, reissued 1990) The Causes of Evolution. Princeton Science Library, Princeton NJ; with an afterword by Egbert G. Leigh Jr.

Julian Huxley (1887–1975)

Huxley, grandson of Darwin's supporter T. H. Huxley, was a firm believer in natural selection while most biologists followed the saltationist school. He travelled widely and collaborated with several leading contributors to classical and population genetics, including Muller, Goldschmidt, Fisher, Wright, Haldane, the chromosome expert Cyril Darlington (1903–1981) and the ecological geneticist Edmund Ford (1901–1988). A lucid writer, Huxley did much to popularise the new rapprochement between Darwinism and classical genetics, and his early book[7] on the subject appeared in several editions. Huxley coined various terms that remain in use in biology, such as *cline* (a subspecies belonging to a particular geographical area). Notably, he invented the label *synthetic theory* for the mature concept of evolution that developed from population genetics.

The Synthetic Theory of Evolution

Dobzhansky: Theoretical Models Reconciled with Field Biology

Another leading contributor to this 'new synthesis' who impressed Huxley was the Ukrainian biologist Theodosius Dobzhansky (1900–1975). During his studies in Leningrad, Dobzhansky had been strongly influenced by work on genetic diversity by Sergei Chetverikov (1880–1959) and others. He emigrated to the United States in 1927 and worked for a time with Morgan. He showed that the theoretical models of population genetics could be applied to empirical field data on the genetic diversity of wild populations, working particularly on regional varieties of *Drosophila*. This reconciliation between population genetics and field biology is captured in his 1937 book,[8] a classic of the synthetic theory of evolution. Like Haldane and Huxley, Dobzhansky was a highly able mathematician but could communicate in lucid, non-mathematical terms. Huxley and Dobzhansky were instrumental in making the synthetic theory accessible to the many biologists for whom the sophisticated mathematics of Fisher and Wright was too difficult.

Mayr: Lamarckian and Saltationist Ideas Eliminated; Evolution Reconciled with Taxonomy

Bernhard Rensch (1900–1990) studied the influence of local environmental factors on the geographical distribution of closely-related species. This work influenced the ornithologist Ernst Mayr (1904–2005). Mayr went on to develop Dobzhansky's

---

[7] Huxley J (1942) Evolution: the Modern Synthesis. Allen & Unwin, London.

[8] Genetics and the Origin of Species. Columbia University Press, New York. (Further editions followed in 1941 and 1951, illustrating the gradual development of the theory.)

studies of wild populations. His first book[9] emphasised the formation of new species in *reproductively isolated* wild populations. He showed that natural selection could explain the whole of evolution, including why genes evolve at the molecular level, and left no place for Lamarckian or saltationist ideas in biology. He also reformulated the species concept, defining a species as a group of interbreeding or potentially interbreeding populations that were reproductively isolated from all other populations. This reconciled the synthetic theory of evolution with taxonomy.

Simpson and de Beer: Synthetic Theory Reconciled with Palaeontology and Embryology

Two years later, the palaeontologist George Gaylord Simpson (1902–1984) showed[10] that the fossil record is consistent with the irregular non-directional pattern of evolution predicted by the synthetic theory. Simpson supplemented Mayr's account of evolution by showing that species can change gradually without the splitting of lineages by reproductive isolation. Like Mayr, he showed that Lamarckian and saltationist models could not explain evolution.

Botany was explicitly assimilated into the synthetic theory by G. Ledyard Stebbins (1906–2000) in 1950.[11]

Thus, within little more than a decade, the mature version of evolutionary theory had become cross-disciplinary, integrating traditional botany and zoology, genetics, ecology, palaeontology and aspects of cell biology. Through the work of Gavin de Beer (1899–1972), it also became consistent with embryology. De Beer worked with Haldane and Huxley during the 1920s and 1930s, particularly on experimental embryology. In a 1930 publication[12] he showed that apparent sudden 'jumps' in the fossil record could be explained by reference to embryo development. A novel feature may develop gradually, by natural selection, in the juvenile form of an animal. Juvenile forms do not usually leave fossils. If at some later stage this species reached sexual maturity while retaining many juvenile features – a well-known phenomenon called *neoteny* – then fossils of the animal would show the novelty, with no apparent precedent. Thus, De Beer's embryological work corroborated Simpson's argument that the fossil record was consistent with Darwinian natural selection. This refuted Goldschmidt's saltationist claim that 'macroevolutionary' changes in chromosome patterns were more important in evolution than the 'microevolutionary' changes of natural selection.

---

[9] (1942) Systematics and the Origin of Species. Harvard University Press, Cambridge, MA.

[10] (1944) Mode and Tempo in Evolution. Columbia University Press, New York.

[11] (1950) Variation and Evolution in Plants. Columbia University Press, New York.

[12] Embryos and Evolution. Oxford University Press, Oxford.

Prior to the work of de Beer and Simpson, the fossil record had seemed to support Goldschmidt's latter-day saltationist view. The proponents of the synthetic theory believed de Beer and dismissed Goldschmidt's idea as untenable, just as Morgan's team had dismissed his critique of the crossing-over hypothesis. However, late 20th century advances in the molecular biology of embryo development revealed a set of developmental 'master genes'. If one of these 'master genes' undergoes a mutation, then the course of development of the embryo may be radically altered. This discovery resurrected Goldschmidt's idea, though in an unanticipated form.

The Main Principles of the Synthetic Theory

The main tenets of the synthetic theory in the early 1940s were as follows.

1. Evolution is gradual (*Darwin*). Small genetic changes and recombinations occur (*Morgan*); the phenotypic consequences are then subject to natural selection.
2. Selection is the main mechanism of change; even slight advantages are important when sustained through several generations (*Darwin*).
3. Genetic drift may be important in evolution (*Wright, Dobzhansky*) or may be of little significance (*Fisher*).
4. Discontinuities amongst species arise through reproductive isolation (*Mayr*), and possibly through neoteny (*de Beer*).
5. The genetic diversity of natural populations is a key factor in evolution (*Fisher, Dobzhansky*). Natural selection is a strong evolutionary force in wild populations (*Haldane, Dobzhansky*).
6. The fossil record can be explained by extrapolating from micro- to macro-evolution (*Simpson*) (Fig. 14.1).

Biochemistry is Reconciled with the Synthetic Theory

Biochemistry emerged from 19th century developments in organic chemistry, cell biology and experimental physiology (Chapter 9). During the early 20th century, biochemists were mainly concerned with the nature and activities of enzymes and their role in metabolism. Before the Second World War they had elucidated the major metabolic pathways that we described in Chapter 4 of *About Life*.

The early biochemists realised that enzymes are highly specific. They are crucial for maintaining cell life and thus the normal physiology of the organism. Genes, like enzymes, are highly specific and essential for normal physiology. Therefore, like enzymes, they must surely be very complex molecules. As we have seen, Sewell Wright speculated (presciently) that genes act by controlling enzymes; Richard Goldschmidt believed that genes *are* enzymes. The most conservative hypothesis seemed to be that each gene is responsible for producing a single

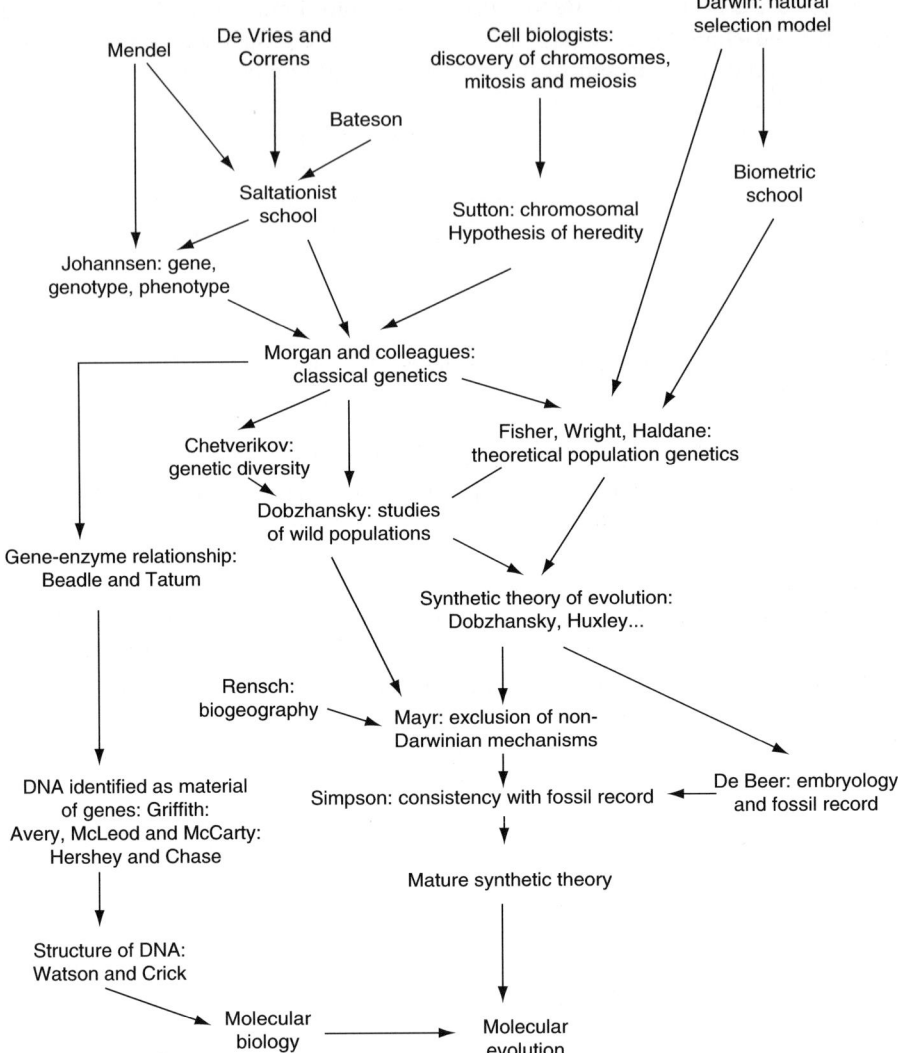

Fig. 14.1 Main ingredients of the synthetic theory of evolution. This summary flow diagram shows, in roughly chronological order, how many disparate ideas were gradually combined to create the synthetic theory of evolution. The main lines of development can be traced to Darwin in the mid-19th century and to the saltationist school at the start of the 20th (and through them to Mendel). The diagram is extended to incorporate the discovery of DNA structure and the growth of molecular biology, which is discussed in the later part of this chapter

enzyme.[13] George Beadle (1903–1989) and Edward Tatum (1909–1975) found a way of testing this hypothesis experimentally. The techniques available to them were very limited by present-day standards and their work was a masterpiece of experiment design.

The mycologist Bernard Dodge (1872–1960) had shown that Mendelian genetics applied to a lowly fungus, *Neurospora crassa*, which sometimes appears as a red mould on bread. *Neurospora* can be grown on a 'minimal medium' containing only salts and very basic nutrients. It is haploid, like an animal or plant gamete, i.e. it has only a 'half chromosome number' (Chapters 10 and 13). Like many fungi, it undergoes a form of sexual reproduction in which parts of two individuals fuse to produce a cell containing paired homologous chromosomes (i.e. a diploid cell). The diploid cell then divides to produce spores, each of which is haploid. The spores are very numerous, resistant and durable. Under the right conditions they grow into new haploid *Neurospora*.

Realising that this fungus could be used for genetic experiments, Dodge recommended it to Morgan, but he ignored it and continued to work on *Drosophila*. Beadle and Tatum, on the other hand, saw how they might use *Neurospora* to test their hypothesis. They focused on the process by which the fungus manufactures an amino acid, arginine.

Morgan's colleague Muller had discovered that harmful radiation such as X-rays can cause mutations. Beadle and Tatum irradiated *Neurospora* and waited for spores to form. Most of these spores grew on minimal medium, but some did not. These 'failed spores' were then transferred to the same minimal medium supplemented with arginine, whereupon some of them grew into apparently normal *Neurospora*. Clearly, these selected spores were undamaged *except* that they lacked an enzyme needed to make arginine. Further experiments identified seven distinct arginine-requiring mutants. Beadle and Tatum concluded that at least seven enzymes were required to make arginine, and each of these was produced by a gene that could be inactivated by mutagenic X-rays.

This study effectively linked genetics – and therefore the synthetic theory of evolution – to biochemistry. Over the decades that followed, the gene-enzyme relationship was corroborated but made more general: every *protein* (strictly, every *polypeptide*), not just every enzyme, is encoded in a gene. Because many proteins, including the enzymes involved in metabolism, are found in more or less all organisms, it therefore follows that some genes are 'universal'. However, the amino acid sequence of any protein differs from one species to another. Therefore, the DNA sequence in the gene encoding that protein also differs. The closer the evolutionary relationship between the species, the more alike the sequences are. Nowadays, this can be demonstrated experimentally using gene sequencing methods.

---

[13] In 1902, Archibald Garrod suggested that some diseases might result from gene defects. He called them 'inborn errors of metabolism', a phrase that remains in use, and wrote an article about the specific example of alkaptonuria: Garrod AE (1902) The incidence of alkaptonuria: a study of chemical individuality. Lancet ii:1616–1620. Bateson was impressed by Garrod's suggestion, but Morgan ignored it; the hypothesis could not be tested experimentally at the time. The later work of Beadle and Tatum helped to show that Garrod was correct.

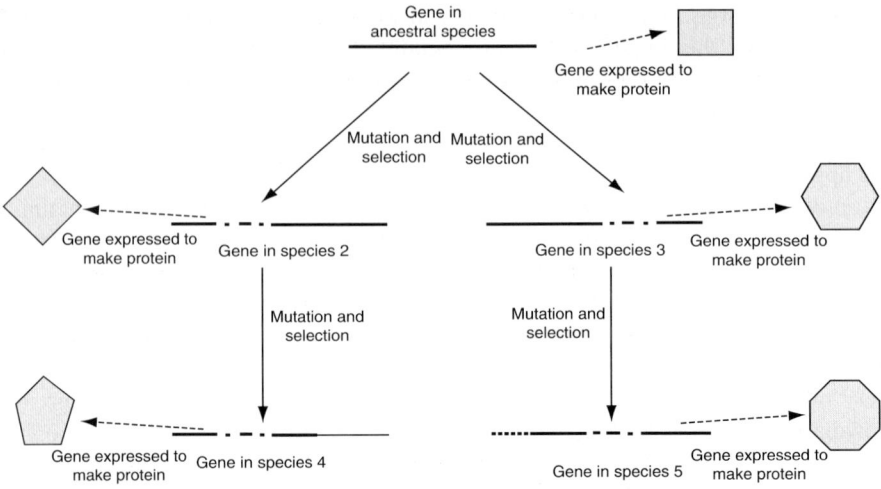

Fig. 14.2 Schematic illustration of molecular evolution. The diagram shows a simple, hypothetical evolutionary sequence in two branches, one giving rise to species 2 and 4 and the other to species 3 and 5. The dense horizontal lines represent a gene that undergoes mutations, producing different *alleles* in the different species. The gene encodes a protein, which is represented by a geometrical figure. As mutations produce different alleles, the protein is changed (the shape of the geometrical figure changes). These changes in the protein are the basis of differences in the *phenotype* of the organism. Natural selection acts on the phenotype

Proteins evolve as genes evolve, and mutations cause changes in protein sequences. These underlie the production of phenotypic variants, which are then subject to natural selection (Fig. 14.2).

Bacterial Genetics

In the early 20th century, bacteria were regarded as quite separate from other organisms. No one had considered that the theories of genetics and evolution might apply to them. But in 1928, the medical bacteriologist Frederick Griffith (1879–1941) performed a classic series of experiments with *Diplococcus pneumoniae*, an organism that causes pneumonia in humans and in other mammals including mice.

Griffith used two strains of *Diplococcus*, one virulent and one harmless. The virulent ones have capsules, the harmless ones do not. If the virulent strain was injected into mice, the mice died; but if the bacteria were heated first, they lost their virulence and the mice lived. Of course, the harmless non-capsulated strain did not kill the mice. However, if the heat-inactivated virulent bacteria were mixed with the harmless strain and the mixture was injected, the mice died. This extraordinary result was reproducible. Somehow, the ability to make a capsule and therefore to become virulent had been transferred from the dead capsulated bacteria to the live uncapsulated ones.

Clearly, a chemical substance, a 'virulence factor', must have been transferred from the dead bacteria to the live ones. The transforming substance behaved like a gene, with the capsule as its product. Could the transforming substance be extracted, used to transform uncapsulated bacteria in a test-tube, and then identified? Would it prove to be the material of which genes are made? Several bacteriologists were excited by the possibility, but geneticists mostly ignored it - they showed no interest in bacteria.

More than a decade passed before Oswald Avery (1877–1955) and his colleagues Colin McLeod (1909–1972) and Maclyn McCarty (1911–2005) succeeded in isolating the transforming substance.[14] It was extremely active (one part in 600,000,000 transformed unencapsulated *Diplococcus* in a test tube) and had an enormously high molecular weight. It was not protein so it was not an enzyme, nor was it lipid. It proved to be DNA.

At first, this discovery evoked widespread scepticism. DNA is a chemically *simple* molecule despite its great molecular size; the complicated tasks performed by a gene surely required a *complicated* molecule, almost certainly a protein. DNA was known to be present in chromosomes, but most people believed that it only provided a skeletal framework to which the proteins, the 'real' substance of genes, were attached. The ensuing debate continued into the early 1950s: were genes made of protein or were they, after all, DNA?

In 1952, Alfred Hershey (1908–1997) and Martha Chase (1927–2003) obtained convincing evidence that the genetic material *is* DNA. They used a new technique, radioisotope labelling,[15] to study a bacteriophage. A bacteriophage is a virus that infects bacteria. It usurps the protein-making machinery of the target bacterium, making it produce bacteriophage proteins instead of bacterial proteins. In other words, it puts its own genes into the bacterium. Hershey and Chase labelled the protein coat of their bacteriophage with one isotope and the DNA with another; the DNA label entered the infected bacterium while the protein label stayed on the outside. The conclusion was clear: bacteriophage genes, and by implication *all* genes, are DNA not protein.

These and subsequent studies showed that bacteria have genes. They evolve, just as eukaryotes do, and thus the synthetic theory became applicable to prokaryotes.

The Molecular Basis of Heredity and Evolution

One of the most famous of all scientific papers, the first description of the double-helix structure of DNA, is one page long[16] and is written in beautifully under-stated prose. It begins: '*We wish to suggest a structure for the salt of deoxyribose*

---

[14] Avery OT, MacLeod CM, McCarty M (1944) Studies on the nature of the substance inducing transformation of pneumococcal types. *J Exp Med* 79:137–158.

[15] Hershey AD, Chase M (1952) Independent functions of viral protein and nucleic acid in growth of bacteriophage. *J Gen Microbiol* 36:39–56.

[16] Watson JD, Crick FHC (1953) A structure for deoxyribose nucleic acid. *Nature* 171:737.

nucleic acid (D.N.A.). This structure has novel features which are of considerable biological interest.' The penultimate paragraph consists of the immortal sentence '*It has not escaped our notice that the specific pairing we have postulated immediately suggests a possible copying mechanism for the genetic material*'. That is a crucial feature of the double-helix model – it allows DNA to be replicated exactly. So if genes are made of DNA, the model can account for the exact replication of genes.

Because of the Beadle-Tatum experiments and their successors, it was accepted that genes specify the manufacture of proteins (including enzymes). The major challenges that remained after the Watson-Crick model of DNA were to explain (1) how the DNA molecule serves as a store of genetic information, (2) how it can change so as to account for mutations, (3) how the genetic information can be translated to make proteins. These challenges were met during the 1950s and 1960s; we gave the answers in Chapter 7 of *About Life*. Among the many contributors to these advances, which created the discipline of *molecular biology*, Francis Crick (1916–2004) remained prominent. We now know that:

1. DNA serves as a store of genetic information by virtue of specific sequences of nucleotide bases (A, G, T, C), just as the information in written English words depends on the sequences in which the letters are arranged.
2. A sequence can be changed by many different mechanisms, most of which have been elucidated during the past quarter century (Chapter 11 of *About Life*).
3. The *genetic code* was cracked, and the outline of the protein synthesis mechanism was understood by the mid-1960s.

Molecular biology has added a 'chemical dimension' to the synthetic theory of evolution. Genes are sequences of DNA; mutations are changes in that sequence; heredity is underpinned by the exact copying of DNA sequences. The universality of the genetic code dramatically demonstrates the unity of all life from prokaryotes to humans, and it points to an ultimate common ancestor for all extant species. Evolution has a molecular basis: changes in DNA caused by the accumulation of mutations over time. Modern DNA sequencing and other techniques make it possible to compare the genomes of related species and to understand more clearly how they have evolved.

Since the early 1980s, molecular biologists have also obtained insights into the mechanisms underlying embryo development. Among other things, these insights have shown that the diversification of animal forms during evolution has been underpinned by the actions of a small set of proteins, the 'developmental toolkit', which regulates the development of all animal embryos. Comparative developmental biology has now reinvigorated the old relationship between embryology and evolutionary theory (Chapter 10) in the form of a new discipline dubbed 'evo-devo'. The genome, a product of evolution, is a 'recipe' for the organism; it is increasingly being seen as a *developmental program*. Ernst Haeckel would have been delighted.

The Modern Theory of Evolution

We have devoted the three chapters to the theory of evolution. Like any good scientific theory, it has undergone serial changes and continues to do so. It has all the characteristics of a scientific theory that we listed in Chapter 1 – it is naturalistic, mechanistic, value-neutral, general, reductionist, comprehensive, abstract, simple, logically coherent, 'public', impersonal and inherently progressive. These qualities have accrued gradually: before Darwin it was not mechanistic; before the development of population genetics it was concrete rather than abstract; before the synthetic theory matured it was not fully coherent. There is now a clear consensus among biologists: *...all organisms are the result of descent with modification from a common ancestor; modification results from changes in the genetic material, the DNA, and these are heritable; and natural selection is the main determinant of which variants survive at any time in history.* Gene modification (mutation) proposes; natural selection disposes.

At root, evolution is a theory explaining the diversity of life. It also entails the *continuity* of life, so it is consistent with cell theory but inconsistent with spontaneous generation. However, it is also a 'unifying' theory in two distinct senses. First, along with cell biology, it shows that all life is unified, most obviously and dramatically in the universality of the genetic code. Second, as we have seen in the course of this chapter, it unifies biology as a science. It interconnects all the several disciplines that come under the general heading 'biology'. For these reasons, it is indispensable.

As remarked earlier, the theory of evolution is also indispensable because it provides the only known means by which we can explain 'purposiveness' mechanistically, dispensing with 'entelechy' and the last remaining shreds of Aristotelianism, thereby making biology truly scientific. We devote the next chapter to this crucially important topic.

Chapter 15
The Problem of Purpose[1]

Living organisms act purposefully, and their individual parts – organs, cells, organelles, molecules – fulfil purposes for the whole. Those purposes 'come from within'; animals, for example, seek food and mates *for themselves*. In contrast, the purposes of technological products such as drawing pins, hat-stands and washing machines 'come from outside'; they are defined by their makers and users. A washing machine does not wash clothes *for itself*.

Ever since the Scientific Revolution it has been agreed that the inanimate world of rocks, rivers, stars, clouds etc. does not act purposefully. It is to be understood in mechanistic not teleological terms. An inanimate object is not *for* anything or anyone; whatever it does is a consequence of antecedent causes. That implies a basic difference between biology on the one hand and physics and chemistry on the other. If biology is to be wholly compatible with physics, that difference needs to be resolved. We must be able to make complete *mechanistic* sense of purpose in biology.

But that is not a straightforward matter; 'purpose' is a slippery concept. We can propose (for example) that the *purpose* of a flower colour is to attract pollinating insects, but the proposal is not testable; it is not a hypothesis. We may be able to show experimentally that if the colour is changed then the flower fails to attract the insects, but that only demonstrates an *effect* of the colour. Can we legitimately equate 'effect' with 'purpose'? We argue in this chapter that we cannot.

Another difficulty with 'purpose' statements is that they seem to invert the chronology of causal relationships. Consider a trivial example: 'the purpose of my washing machine is to enable me to wash clothes'. The 'teleological cause' of the washing machine (i.e. its purpose, to wash clothes), comes *after*, not *before*, the 'teleological effect' (i.e. the existence of the machine). I cannot wash clothes at time **t** unless I have the washing machine at a time *prior* to **t**. Contrast that with: 'gravity caused the apple to fall from the tree to the ground'. The cause (gravity) was in operation *before* the effect (the fall of the apple). If we are to make

---

[1] Much of the argument of this chapter derives from Agutter PS, Wheatley DN (1999) On the "problem of purpose" in biology and our acceptance of the Darwinian theory of natural selection. *Found Sci* **4:**3–23. There is a summary account in Agutter PS, Wheatley DN (1997) Teleology in Biology. *Biologist* 44:432.

mechanistic sense of 'purpose' statements, we need to account for this apparent reversal of chronology.

Aristotle had no difficulty with teleological statements. He explained biological phenomena, particularly embryo development, in terms of entelechy. The main drawback was that it 'fixed' his explanations, placing them beyond critical scrutiny and precluding the *progression* of knowledge that is characteristic of science. We considered examples in Chapters 5 and 6. Aristotle's views were challenged during and after the Scientific Revolution and attempts were made to evade the 'problem of purpose', but as we have seen (Chapters 8 and 9), none of them proved satisfactory until the synthetic theory of evolution matured.

The mature theory of evolution is mechanistic, not teleological; it does not acknowledge purpose or design (Chapter 12). How, then, can the ineluctable purposiveness of organisms and their parts be reconciled with a theory that *denies* design and purpose? More particularly, how can it be explained in terms of that very theory?

Artefacts and Organisms: Another View of the Living State

Organisms, like technological artefacts, seem 'designed': the functions of their parts can only be understood in terms of the purposes of the whole. But we are unlikely to confuse an artefact with an organism. Nobody in a sober condition would believe a hat-stand to be alive, or mistake a live dog for a mechanical one. How do we know the difference? What tells us that X is an organism and Y is an artefact?

As we noted at the start of this chapter, an organism has its own goals, ultimately those of survival and transmission of its genes, while an artefact serves its designer's or user's purposes. But that does not enable us to distinguish organism from artefact; we cannot observe the 'source' of a purpose. Nevertheless, there is a useful related point: the parts of an artefact are made and set in place by the designer or manufacturer, while the parts of an organism are made *by, within and for the organism itself.* That characteristic of organisms was recognised by Immanuel Kant and was emphasised more recently by Robert Rosen. It is part of what we mean when we describe organisms (or cells) as 'autopoietic'. In Rosen's deliberately Aristotelian words, organisms are *closed* with regard to efficient and final causes.

If an object is 'intrinsically endowed with purpose', and all its functional parts are made by and within itself, then that object is an organism not an artefact. This description is consistent with the characterisation of the living state given in our previous book, *About Life*, though it is more abstract. The three-way interdependence of gene expression control, internal state and responsiveness to external stimuli (Chapter 10 in *About Life*) is equivalent to Rosen's 'closure under efficient and final cause'. Every component of a cell originates through the cell's activities, and serves to sustain and modulate those activities so as to ensure the survival and replication of the cell.

The Ambiguity of 'Purpose'

'Purpose' can be used in two apparently distinct ways. One applies to the organism as a whole, the other to its parts. Consider the following sentences:

1. The purpose of a mammal's lungs is to ensure efficient exchange of oxygen and carbon dioxide between the air and the blood stream.
2. Joan's purpose in going to university was to obtain a degree.

We can replace 'purpose' in sentence 1 with 'function' but not with 'goal'. We can replace 'purpose' in sentence 2 with 'goal' but not with 'function'. It would be absurd to speak of the 'goal' of a mammal's lungs, as though lungs had conscious intentions. It would be strange and rather disturbing to speak of a person's 'function' in respect of life choices: if we value the rights of individuals to live as they please, provided they accord with the law, we should not characterise them in terms of 'function'.

For the time being we shall consider function-statements such as (1) above. The treatment we outline below echoes the analyses given by Monod, Mayr and other writers (see the bibliography). We shall turn to goal-statements such as (2) later.

Making Mechanistic Sense of Function-Statements

To avoid the ambiguity of 'purpose', let us substitute 'function' for 'purpose' in (1), above. The substitution preserves the 'chronological inversion' of function statements: the 'teleological cause' (function) of lung-ownership (ensuring efficient respiratory gas exchange) succeeds rather than precedes the 'teleological effect' (existence of the lungs).

When we say that 'the function of mammalian lungs is to ensure efficient exchange of oxygen and carbon dioxide between the air and the blood stream', we mean at least the following:

(a) Mammals do, as a matter of fact, have lungs.
(b) Lungs, when they work properly, do ensure efficient exchange of oxygen and carbon dioxide between the air and the blood stream.

Strictly speaking, we ought to define 'air', 'blood stream' and 'efficient' more precisely, but for the present argument we only need to know that we *can* make exact sense of these terms.[2] Statements (a) and (b) could initiate any number of

---

[2] The 'air' is the gas phase in the lung alveoli, which differs in composition from an average atmospheric sample. The 'blood stream' here denotes the contents of the pulmonary capillaries, which are juxtaposed with the alveoli. 'Efficiency' can be reckoned in terms of the amounts of oxygen and carbon dioxide that pass through the total lung alveolar surface or through a square metre of that surface in unit time. These details are scientifically important but they would obscure the present discussion without adding anything useful to it.

lines of experimental inquiry: lung development from early embryo stages, its genetic basis, adult lung structure, the mechanism of gas movement between alveolar space and lung capillaries, the control of respiratory movements, and so on. But (a) and (b) do not add up to the notion of 'function'. Something else is needed.

Another 'obvious' pair of proposition, (c) and (d), can be introduced. Together with (a) and (b) they seem to complete the analysis of 'function':

(c) Nothing other than lungs ensures efficient respiratory gas exchange in mammals.
(d) Unless efficient respiratory gas exchange is ensured, the mammal will not survive to reproductive age and will therefore not leave offspring.

Proposition (d) is a composite (survival plus leaving offspring), but it will suffice. Jointly, the four propositions (a–d) assert that if mammals are to survive and reproduce, they need to have lungs, which are their only way of ensuring respiratory gas exchange. (Of course, this is not to say that possession of lungs is a *sufficient* condition for survival and reproduction, just that it is a *necessary* one.) 'The function of mammalian lungs is to ensure efficient exchange of oxygen and carbon dioxide between the air and the blood stream' summarises this joint assertion. Let us now examine (a–d) in turn.

(a) Is a general statement justified by induction from observations on all mammals. It is an *empirical generalisation*.
(b) Is a fact of experimental physiology, another empirical generalisation. However, it makes an explicitly *causal* statement, which is mechanistic not teleological: possession of lungs at time **t** is the precondition for efficient gas exchange at **t** and at *subsequent* times, not earlier times.
(c) Is also an empirical generalisation, though its significance is enriched by comparative biological studies.

Clearly (a–c) do not account for the temporal inversion in the function-statement, so the appearance of 'cause succeeding effect' must somehow reside in (d), or in the combination of (d) with (a–c). Statement (d) asserts that survival and reproduction are, among other things, consequences of efficient respiratory gas exchange. Therefore, according to the modern theory of evolution, the device that ensures respiratory gas exchange (the lung) has been perfected by natural selection, making it a *consequence* of the requirements for survival and reproduction. The language of (d) reveals its roots in evolutionary theory. It can only be justified by reference to that theory.

This leads to the crucial implication: *our theory of evolution enables us to paraphrase any function-statement in biology in purely mechanistic terms*. The 'temporal inversion' problem is solved, the need to invoke an 'entelechy' (or a vital force) to explain why parts of organisms are purposive is obviated, and an apparent barrier between physics and biology, the legitimacy of teleological statements, is surmounted. If it were not for Darwin and his intellectual successors we would still be troubled by temporal inversion, arguing about vitalism and haunted by the ghost of Aristotle.

Further Analysis of Function-Statements

A possible objection to the foregoing analysis is that 'function' only means 'effect'. Have we demonstrated anything more than 'an *effect* of mammalian lungs is to ensure efficient exchange of oxygen and carbon dioxide between the air and the blood stream'?

It is easier to analyse this objection if we use a different example. Compare the following two statements:

(i) A function of chlorophyll in plant leaves is to enable photosynthesis to take place.
(ii) A function of chlorophyll in plant leaves is to make them green.

Statement (i) is unexceptionable, though it needs more detailed explanation; but (ii) is a weak joke or a young child's naïveté. If we replace 'a function' with 'an effect', then *both* statements become unexceptionable (though not very useful). Therefore, 'function' is *not* merely another word for 'effect'.

The distinction is quite easy to grasp. By and large, the colour of leaves (or of chlorophyll) is irrelevant to the survival and reproduction of plants, but the capacity to carry out photosynthesis is crucial. 'Function', we might say, is 'an effect that greatly increases the likelihood of survival and reproduction'.

Another possible objection to our analysis is that some parts of organisms are apparently functionless.[3] That seems to cast doubt on Rosen's (Kantian) idea that all (structural and functional) components of an organism are 'closed under efficient and final causes'. The giant panda is a species of bear and has a carnivore's dentition and digestive tract, but because of the environment to which it has been confined in the wild for many generations, its diet is almost exclusively bamboo. The giant panda's nutrition is so inefficient – it has to eat enormous quantities of bamboo on a weight-for-weight basis to survive, and most of the ingested material is egested virtually unchanged – that the species is likely to become extinct.

Why did a species that evolved as a carnivore adopt the diet of a herbivore? A biologist's answer would hinge on adaptation to a particular terrain and climate, the effects of predators, parasites, competitors for resources and so on, and the heritability of core behavioural traits (including those affecting food choices). Detailed answers may outline the evolutionary events that led to the present state of affairs, but the outline would involve speculation; it would have the character of a 'Just So' story rather than a thoroughly-researched scientific account. That is inevitable because of the historical character of evolutionary theory (Chapter 16).

Another question arises from the giant panda example: is the animal really 'evolutionarily designed' as a carnivore? If not, then the inference that it is likely to become extinct because of its nutritional peculiarity loses force. The dentition and the

---

[3] The usual example quoted here is the human appendix, which is reputed to serve only as a site of infection, resulting in excruciatingly painful and potentially life-threatening disease. In fact, the appendix is a significant piece of lymphoid tissue, and although the rest of the body's lymphoid tissue can compensate for appendectomy (just as it suffices to cope with removal of the spleen), it is wrong to deem the appendix 'functionless'.

nature (particularly the length) of the digestive tract provide good circumstantial evidence,[4] but hypotheses should make critically testable predictions. One clear prediction of the hypothesis 'the giant panda is descended from carnivores and retains a carnivore's physiology' is that the levels of certain enzymes in the panda's body (arginase in its liver, for example) will accord with carnivore rather than herbivore standards. As far as we know, this prediction has never been tested experimentally; but it could be, quite easily. The point is that an evolutionary argument arising from a question of 'misplaced function' leads to critically testable hypotheses and therefore, potentially, to new lines of scientific inquiry and new information. Legitimate function-statements in biology can therefore be scientifically productive because we can paraphrase them in mechanistic terms, thanks to the theory of evolution.

We saw in Chapter 14 that the theory of evolution underpins and unifies the whole of modern biology. Now we can add another very important point: without the theory of evolution, we could not make mechanistic sense of the teleological propositions and questions that are inescapable in biology.

Goal-Seeking Behaviour

We noted earlier that 'purpose' can mean 'goal' rather than 'function'. We now turn to this alternative usage. To avoid confusing the argument with examples involving conscious human intention, such as 'Joan's purpose in going to university was to obtain a degree' we shall stick to illustrations involving (i) non-human organisms and (ii) artefacts.

In what sense can an *artefact* be said to have a 'goal' or 'target'? Some artefacts (such as guided missiles and thermostatically-controlled heating systems) involve feedback control or servo-mechanisms, and they exhibit goal-seeking behaviour. They compare their outputs to a prescribed 'goal state' and continue to do so until that goal state is attained. Thus, a thermostatically-controlled domestic heating system switches itself on when its sensors are activated (when the ambient is cooler than the set temperature) and switches itself off again when the sensors are deactivated (when the room is warm enough). There are innumerable biological analogies. *Homeostasis* consists largely in the actions of biological servo-mechanisms. For example, temperature control in mammals involves temperature sensors, various effectors for accelerating the generation or the loss of body heat, and a system for integrating the actions of these sensors and effectors. As a whole, this system 'seeks the goal' of maintaining core body temperature within narrow limits (unless the 'setting' is altered, e.g. when our temperature rises during an infection).

---

[4] The ratio between the length of the stretched-out digestive tract and the length of the animal from nose to tail differs between carnivores and herbivores. In carnivores this ratio is around 5–6; in herbivores it can reach 30 or more. That is because the digestion of plant material takes a long time and requires the action of large populations of bacteria that live in parts of the intestine; no such constraint applies to the digestion of meat. The length of the panda's digestive tract in relation to the length of the animal conforms more closely to carnivore than to herbivore standards.

We consider the 'goals' or 'targets' sought and achieved by biological servo-mechanisms to be subservient to the wider goal of ensuring the survival of the organism by the transmission of its genes. Thus, what appears as a 'goal' or 'target' from the point of view of the homeostatic system is a 'function' from the point of view of the organism. Presumably we can accept the use of 'goal/target' in reference to (say) the temperature-control system of the body, but not in reference to a lung: the former is analogous to familiar goal- or target-seeking artefacts such as thermostats and guided missiles. The relationship between 'goal/target' and 'function' in such artefacts is the same as that in organisms; for the occupants of a building, the *function* of the heating system is to maintain the environment at a bearable temperature.

We can extend this understanding of 'goal/target' to other areas of biology. For instance, mammalian blood contains a type of white blood cell called a neutrophil, which, when suitably 'primed', attacks and scavenges foreign bacteria. From the organism's point of view, the *function* of the neutrophil is to attack and devour invading bacteria. From the neutrophil's point of view, an invading bacterium is a *target* and the ingestion and destruction of that target is the cell's *goal*. When an animal exhibits mating behaviour, its *goal* is to mate; but from the point of view of the species (or the population), the *function* of mating behaviour is to ensure the survival of certain alleles of genes for at least one more generation. In biology, therefore, 'function' and 'goal' are two sides of the same coin. Thus, our mechanistic understanding of function constitutes a mechanistic understanding of 'goal'.

The Limits of Meaningful Teleology?

Can we legitimately speak about the 'purpose' of an individual *organism*, or of a *population* or *species*? A biologist's intuitive response is 'no'.[5] To speak of the 'goals/targets' of an individual organism implies a degree of anthropomorphism that biologists find unacceptable. Nevertheless, our argument may enable us to read 'purpose' as 'function' in such cases. Every organism is adapted to its ecosystem so that the energy flux through that part of the system is maximised. Therefore, we could regard the *function* of an organism (or a population) as *the maximisation of energy flow within the ecosystem of which it forms a part*. In other words, we may once again change the viewpoint from the part (organism or population) to the whole (ecosystem), substituting 'function' for 'goal' in the process.

This approach has limits. Scientifically, we cannot extrapolate beyond the ecosystem level. Even if we choose to say that an ecosystem 'functions' within the context of 'Gaia', there is no context in which the Earth's biosphere as a whole

---

[5] One of us recalls being asked after an open public lecture one evening: 'what are blackbirds for?' Questions about the 'purpose' of an organism, or a species, are disconcerting; as biologists we cannot easily talk about them. The speaker was duly disconcerted and the questioner was left unanswered.

could be said to 'function': the notion of 'goal' or 'target' of the biosphere makes no scientific sense. The biosphere is the boundary beyond which our mechanistic parsing of biological function statements cannot pass.

Must we insist that 'scientifically meaningless' is equivalent to 'meaningless'? A positivist would say so. A mechanistic materialist would deny that any consideration of function or purpose has scientific value. But if we avoid such extreme philosophical positions, we should admit the possibility that assertions about the 'purpose of the biosphere' – or of the planet, or of the universe as a whole – may be meaningful, though they are not *scientifically* meaningful.

By abandoning teleological explanations in favour of mechanistic ones, science has grown into by far the largest, most 'objective' and most productive body of knowledge and belief the world has ever known. 'Purpose' is otiose in theories of physics *and* in the theory of evolution, so it makes no scientific sense to speak of the 'purpose' (or 'function') of the Earth, or of the universe. But unless we side with the positivists and regard science as the only arbiter of truth, that need not imply that it makes no sense of any sort.

Chapter 16
The Scientific Status of Biology

Today, physics and biology share all the general features of science that we surveyed in Chapter 1 (naturalism, mechanism, progressiveness, etc.). Everyone now agrees that living matter obeys all and only the laws of physics and chemistry. This consensus was achieved relatively recently; as we have seen, biology did not shake off the last vestiges of Aristotelianism until long after physics had become unassailably mechanistic. But it is clear that modern biology, like physics, is a *science* in the sense outlined in Chapter 1.

That may seem to imply that modern biologists and physicists think in much the same ways, produce knowledge that is similar in kind (allowing for differences in subject matter) and, as students, undergo comparable learning experiences. The data, however, suggest otherwise. If you ask science students 'How does the study of physics contrast with the study of biology?' they will answer along the following lines. Physics is mathematical, abstract, universal in application, not dependent on factual detail (except in specific applications) and concerned with essentially simple systems. Biology, in contrast, is qualitative, concrete, applicable only to terrestrial organisms, heavily burdened with factual detail and concerned with irreducibly complicated systems. When you study biology you have to think concretely and memorise numerous facts. When you study physics you have to master abstract reasoning and mathematical techniques. Physics and biology text-books and examination papers corroborate the distinction. This evidence indicates that physicists and biologists think differently, and that the kinds of knowledge they generate are dissimilar in kind as well as content.

Can Biology Be 'Reduced' to Physics?

Let us suppose that the differences between the two sciences are not fundamental but will evaporate when biology matures further. In other words, biology will ultimately prove 'reducible' to physics. What would such 'reduction' entail? We will consider this question in terms of cell biology and chemistry (regarding chemistry as a 'physical science'). Formally, two conditions would have to be met in order to accomplish the reduction of cell biology to chemistry:

P.S. Agutter, D.N. Wheatley, *Thinking about Life*,
© Springer Science + Business Media B.V. 2008

1. All *concepts* in cell biology would have to be described – completely, with no semantic residue – in the language of chemistry (and physics).
2. All *principles* evoked in cell biology would have to be logically deducible from principles of chemistry (and physics).

It is logically impossible to meet those conditions.[*]

1. Concepts such as 'control' and 'transcription' in cell biology have no equivalents in physics and chemistry. They are borrowed from control engineering. We can, at cost of being very long-winded, describe the *process* or *mechanism* of transcription in the language of chemistry, but no such description captures the *concept*. By analogy: we could describe the interactions between ink, quill pen and parchment in terms of physics and chemistry but fail to capture the essence of Chaucer's *Canterbury Tales* or Goethe's *Faust*.
2. Many arguments in cell biology implicitly or explicitly invoke the theory of evolution (Chapters 14 and 15). The theory of evolution is irreducibly *historical*: the evolutionary past does not determine the present or the present the future (see below). No theory in physics and chemistry is 'historical' in that sense.[1] Therefore, the principles of an argument invoking evolutionary theory cannot be deduced from principles of chemistry and physics.

Why, then, should we not simply accept that the two kinds of science are fundamentally different? There is a *pragmatic* reason: for at least a century, biology has progressed largely (though not entirely) through the application of concepts and methods borrowed from chemistry and physics. Further progress may therefore depend on continued 'borrowing'. But that not a compelling argument. In principle, further progress may equally well *not* depend on continued 'borrowing'.

There is also an *ideological* reason: the 'reduction' of biology to physics is a tenet of positivism (see Chapter 9). Positivism, as we have seen, was an influential philosophy of science in the 19th and early 20th centuries. It was subsequently shown to be flawed but it remains an *ideology* of science. We are taught to believe that physics is basic to every other science, so there is an 'imperative' to 'reduce' biology to it. That, of course, is an article of faith; commitment to it is emotive, not rational, and not everyone subscribes to it.

Therefore, the urge to assimilate biology into the corpus of physics derives from tradition and ideology, not from scientific reasoning, and in any case the aim cannot be achieved. We conclude that the differences between the two kinds of science are indeed fundamental.

Where exactly do those difference lie?

---

[*] See e.g. Polanyi M (1969) Life's irreducible structure. *Science* 160:1308-1312.
[1] Astronomy and cosmology have strong historical themes, and indeed people speak of the birth of death of stars and the evolution of galaxies. But these 'histories' arise from processes that are described and explained by fundamental theories of physics, which are not in themselves historical. In contrast, the fundamental theory of biology, evolution, is *intrinsically* historical.

Matter, Energy, Information and Organisation

Matter

In Chapter 8 we examined 17th–19th century ideas about the distinction between the living and the non-living, emphasising the debate about organic and inorganic matter. Our conclusions were:

1. The elements carbon, hydrogen, oxygen and nitrogen predominate in biological matter. There is much less carbon and scarcely any nitrogen in non-living matter such as clay and sand, but there is plenty of silicon and aluminium, elements that are rare in biology.
2. Most biological molecules are very much bigger than more complicated than those of non-living matter. The inorganic world has nothing remotely comparable to DNA or proteins.
3. Biological molecules have very specific shapes – the 'handedness' discovered by Pasteur – but non-biological molecules generally do not.

These differences are important but they are not *fundamental*. They do not challenge that assertion that the laws of physics and chemistry apply equally to living and non-living matter.

Energy

The concept of 'energy' evolved over many generations of research. It is one of the most abstract ideas in physics, though it is not difficult to grasp. When a person or a machine performs any activity – *work* – the activity entails a 'cost'. Energy is the currency in which that cost is paid. Scientists recognise many different types of energy: heat, chemical energy, electricity, electromagnetic radiation (radio waves, infrared, light, ultraviolet, X-rays and gamma rays), mechanical movement. These modes can be interconverted. An ordinary torch battery uses a chemical reaction (chemical energy) to produce an electric current, which the bulb's filament converts into light and heat. Energy is never lost; it only changes from one form to another, ultimately changing into its lowest-grade form, heat.

A 19th century theory of energy, classical thermodynamics, enables us to measure amounts of energy and to interrelate its different modes. Classical thermodynamics was a by-product of the age of steam engines, an example of the interplay between culture, belief and technology (Chapter 2), but it applies to the whole of living and non-living nature as well as machines. Your body can change the chemical energy from the food you eat into mechanical energy (muscular work), heat, electrical energy (nerve conduction). If you are a firefly, it can turn that energy into light.

Biologists are concerned with the *mechanisms* involved in these interconversions and particularly with the ways in which energy is channelled to perform biologically useful work, while physicists and chemists are less concerned with such fine details. Thus, the physical laws of thermodynamics give us a necessary basis for understanding biological energy transductions, but not a sufficiently detailed account. Again, however, there is no fundamental difference here between the two kinds of science. The laws of thermodynamics are the same no matter whether they are applied to biological or physico-chemical processes. They are simply *used* differently by physicists and biologists.

Information

Our ideas of matter and energy are among the great intellectual achievements of the 19th century. The theory of information is one of the great achievements of the 20th. It was produced not by scientists but by engineers whose interest lay in tele-communications, servo-systems such as thermostats and guided missiles, and later, computers. Biologists were involved in its early development, but it remains prima-rily an engineer's theory.[2]

'Information' is relevant to biology in two distinct ways. First, we speak about *genetic* information. Every organism is born with a set of genetically-encoded instruc-tions (DNA) for assembling, operating and maintaining that organism. Secondly, we speak about *sensory* information. In complex animals such as ourselves, the ears, eyes, nose, tongue and skin give us moment by moment information about the world around us. Without that continual information input we would not survive. In *About Life*, we suggested that the two kinds of information interact at the level of the cell. In complex animals, there is an analogous interaction at the level of the brain.

The theory of communication articulated during the late 1940s[3] was simple, general and quantitative. It proved valuable for communication engineers who deal (for instance) with the capacities of communication channels in 'bits per second', it was instrumental in the development of computer science, it led to useful work on redundancy in natural language and it gave birth to the mathematical study of information. But as the authors of the theory pointed out,[4] it was not concerned with *meaning:*

> ...two messages, one of which is heavily loaded with meaning and the other of which is pure nonsense, can be exactly equivalent, from the present viewpoint, as regards information.

---

[2] This is superbly discussed in the classic work by Norbert Wiener (1894–1964): Wiener N (1961) Cybernetics or Control and Communication in the Animal and the Machine, 2nd ed. Wiley, New York. The earliest paper we can find that on the subject was written by the great 19th century physicist James Clerk Maxwell (1831–1879), but the topic was engineering – specifically, the governor mechanism on steam engines: (1867–1868) On governors. *Proc Roy Soc A* 16:270–283.

[3] Shannon CE, Weaver W (1949) The Mathematical Theory of Communication. University of Illinois Press, Urbana, IL.

[4] Weaver, W. in Shannon and Weaver, *op. cit.*

It is this, undoubtedly, that Shannon means when he says that "the semantic aspects of communication are irrelevant to the engineering aspects".

The theory has been applied to biological information, but it is not adequate for that purpose because biological information has *semantic* content.[5]

Because the ideas of information in biology do not apply to the non-living world, they seem to imply a real distinction between the biological and the physical sciences. Although biological information always has a specific physico-chemical basis and its transmission can be described mechanistically in physico-chemical terms, there is a genuine conceptual barrier here.

Organisation

Although nothing in biology transgresses the laws of chemistry and physics, organisms could not be *predicted* from those sciences. Words such as 'organisation', 'functional', 'regulatory', 'information' and 'stimulus' are not found in the vocabularies of chemistry and physics.[6] They testify to a whole that is greater than the sum of its parts. *The complex[7] organisation of the cell as a whole is the essence of the living state.* Whilst the physics and chemistry of the molecular parts suffice to explain (in principle) the mechanisms underlying that complex whole, they do not enable us to understand or predict a living entity. By analogy: a detailed knowledge of the chemistry of fibroin, the protein that makes up silk, is not sufficient for us to define a shirt.

Summary

- Ideas of matter and energy differ in detail, but not fundamentally, between the two kinds of science.
- In contrast, the concept of 'information' is highly relevant to biology but alien to the physical sciences. Specific instances of information storage and transmission

---

[5] Maynard Smith J (2000) The concept of information in biology. *Philos Sci* 67:177–194.

[6] Lloyd Demetrius points out that organisation in physical systems is based on thermal laws, organisation in biological systems on temporal laws.

[7] The word 'complex' merits comment. A cell is obviously *complicated* in the sense that it consists of millions of different, often very big, molecules precisely arranged in space and time, but it may also be 'complex' in the modern mathematical sense. Cells are stable (robust) to perturbation, they have emergent properties and they are hierarchically organised. These are the main features of systems that exhibit what Stuart Kauffman calls *self-organising complexity*. Kauffman has shown that such systems have a number of stable states and can progress from one state to another, just as eukaryotic cells do when they are dividing (the 'cell cycle'). Self-organising complexity is exhibited by systems 'on the edge of chaos': not strictly predictable but not quite chaotic either. We discussed this topic in *About Life*.

in biology can be explicated in physico-chemical terms but the concept cannot be 'reduced' to physics and chemistry.

- The organisation of biological matter is essential for the living state and although it is entirely consistent with the laws of physics and chemistry, it cannot be predicted from those sciences.

These considerations throw some light on the differences between the two kinds of science, but they do not give us a complete explanation.

Theories in Physics and Biology

A scientific theory is an organised body of knowledge. It is founded on 'articles of faith' called *postulates*. For example, a postulate of relativity theory is that the velocity of light *in vacuo* cannot be exceeded. A theory contains *laws* or *principles*, some of which relate abstract terms to measurable variables, while others are empirical generalisations. Thus, Newton's law of gravity deploys an abstract term, the universal gravitational constant, to relate the masses of two objects and the distance between them to the force of by which they attract each other. A theory provides the context or language in which hypotheses are formulated, experiments are designed and data are judged relevant or irrelevant. It is our way of understanding particular aspects of the observable world scientifically (see the appendix for further discussion).

Two theories underpin modern physics: quantum theory and relativity theory, both of which are mathematical. Two quite different theories underpin modern biology: cell theory and evolutionary theory, both of which are phenomenological. The two theories of physics were founded in the early 20th century, the two theories of biology in the mid-19th. A major challenge in modern physics is to reconcile relativity with quantum theory, but in biology there is no analogous challenge; cell theory and evolutionary theory are wholly compatible and have been since at least the 1940s. But physics was unequivocally scientific before either quantum theory or relativity theory was conceived; remove either or both and physics would still be a science. In contrast, the scientific status of biology rests entirely on the twin pillars of cell theory and evolutionary theory. Remove either or both and biology would no longer be a science.[8]

---

[8] Those who wish to speculate about so-called 'intelligent design' must therefore realise that in so doing they *necessarily* generate a non-scientific 'biology'. To deny evolution is neither more nor less sensible than to deny the existence of cells. 'Creation science' is inescapably an oxymoron. Readers who are tempted by such notions may be advised to read Dobzhansky's famous essay 'Nothing in biology makes sense except in the light of evolution' (published (1973) in Am Biol Teacher 35:125–129 and now available online at http://people.delphiforums.com/lordorman/light. htm; accessed 7 March 2008). Dobzhansky, a man of deep Christian faith, showed that 'creationism' is as harmful and insulting to religion as it is to science.

Relativity theory concerns very large objects, distances and velocities and is irrelevant to biology. Quantum theory is *indirectly* relevant to biology because it explains the chemical behaviour of atoms and molecules, but biologists very seldom evoke quantum theory in their work. Neither cell theory nor evolutionary theory is relevant to physics or chemistry. Since theories represent the ways in which we think about particular areas of science, it follows that physicists and biologists think in fundamentally different ways.

Let us now consider the ways in which cell and evolutionary theory make biology distinctive.

Cell Theory

The postulates of cell theory (see Chapter 9) may be stated as follows:

- A cell is the smallest unit capable of independent life.
- All organisms consist of one or more cells and their products.
- Cells are produced only from other cells.

These postulates entail the *unity* and *continuity* of life. The first two postulates together exclude viruses from the set of all organisms (cf. *About Life*, Chapter 10).

Since its inception during the mid-19th century, cell theory has grown and developed to assimilate embryology, biochemistry and, more recently, molecular biology. Its range and content are covered in Chapters 2–9 of *About Life*. It contains concepts of structure (e.g. membranes, cytoskeleton), which combine data from microscopy and biochemistry; of metabolism (e.g. metabolic pathway); of responses to external stimuli (e.g. receptor, protein kinase, transcription factor); of DNA replication and repair; of gene expression and its control; and of cell division, differentiation and apoptosis. Like all viable theories it continually grows and develops.

As a whole, cell theory addresses the composition and organisation of living matter and the ways in which organisms obtain, deploy and interconvert different forms of energy. In other words, it incorporates all the ways in which ideas of matter, energy and organisation are distinctive in biology. It also accounts for the processing of both genetic and environmental information. On its own, however, it does not account for the *semantic content* of such information. For that, the theory of evolution is also required.

Evolutionary Theory

At least two postulates underpin the theory of evolution:

- All organisms extant at time **t** are descended from one or more organisms extant at a time prior to **t**.
- All organisms are ultimately descended from the same common ancestor.

The first postulate logically entails the continuity of all life and allows the possibility of diversification. The second specifies a limitation in time: the process of evolution has not been of infinite duration. It also implies the observed universality of the genetic code and the commonality of biochemical components and processes among all organisms. The two postulates together indicate that the theory of evolution cannot apply to the origin of life.

The principles or laws of the theory include the principle of natural selection, the laws of classical genetics, the principles of population genetics and the central dogma of molecular biology. As we saw in Chapters 12–14, the theory continually grows and develops. It overlaps with cell theory in the domains of biochemistry and molecular biology and, increasingly, in 'evo-devo' (Chapter 14).

How does the theory of evolution account for the semantic content of biological information? According to Maynard Smith, the answer is that natural selection causes information to accumulate in genes. This information is copied into daughter cells and is transcribed and translated, or 'decoded', to produce the cell's proteins (gene expression). The translation apparatus constitutes a set of channel conditions for receiving and decoding messages transcribed from the genome. Both replication and gene expression are *unidirectional* processes, in contrast to most information transmission systems contrived by human technology. The genetic code is *arbitrary* ('symbolic'), since no necessary connection between signal and meaning is implicit in the relevant chemistry. It is also *intentional*, since the choice of form is constrained by the message: the genome has a causal role, which is conferred by natural selection. Thus, genetic information has semantic content. Similar arguments apply to sensory information (environmental stimuli).

The Historical Character of Evolutionary Theory

The theory of evolution explains phenomena only retrospectively, by reference to accidental or contingent events in the past: random mutations, unexpected changes in the environment such as volcanic eruptions. It cannot predict future changes in organisms because accidental, contingent, random, unexpected events are by definition unpredictable, and such events will determine future evolutionary developments.

Historical discourse and scientific discourse are quite different. The historian's task is to construct a credible account of past events from incomplete records. History is not reproducible; there are no general laws, no ways of constructing simple models for simulation or laboratory experiment. The present is just one of an unlimited number of possible outcomes of the past, and it contains the potential for an indefinite range of possible futures. History is particular and does not lend itself to generalisations.

Scientists study events that are reproducible, can be modelled, can usually be simulated on the computer or in the laboratory, are governed by general laws and can be explained mechanistically. Such events are characteristic of cells and organisms as well as inanimate physical and chemical systems. The development of an embryo and the fate of a population of cells in an adult animal are programmed. If we perturb the system, for example by altering or knocking out a gene, then we can make rational predictions about the resulting developmental or physiological abnormalities. Cell biology, physiology and embryology are science, not history.

Ecosystem development is another matter. There are general patterns, but detailed predictions about future states cannot be made. If you watched a time-lapse film of a developing embryo and the film were interrupted, you would be able to say more or less exactly what happened next. But if a time-lapse film of an ecosystem were made over (say) a 50-year period, then if the film were interrupted you could only *guess* what happened next. If details of soil type, climate, component species and their interrelationships and so on were known, then a broad idea of the history of an ecosystem could be outlined by an expert. Understanding of food chains, energy flow and interacting population dynamics would reveal patterns. But no one, irrespective of expertise, could predict the behaviour of an ecosystem in detail over an extended period. The actual history would depend too much on accidents and contingencies: floods, forest fires, pestilence, mutations, unexpected emigrations and immigrations of species, climate change and so on. The story revealed by the time-lapse film would be peculiar to that ecosystem over that period of history. It would not be generalisable. Therefore, ecosystem development has more of the character of history than of science – though the methods and theories of ecologists are unquestionably scientific.

Suppose the ecosystem were the whole Earth. Suppose we consider not just a 50-year interval but the entire history of life on the planet. Clearly we are dealing with a *historical* subject, no matter how scientific our methods, our theory and our style of discourse. The study of biological evolution is therefore more akin to historiography than to science as we usually conceive it.

Yet that is not wholly true. Evolution is not fraught with the motivations and personalities of actors, as human history is, nor is the evidence inherently biased. And we have already seen that evolution *does* have the character of a scientific theory. It is historical because neither mutations nor environmental changes are predictable or governed by general laws; but it is scientific because (a) the relationship between genotype and phenotype is in principle deterministic, and (b) selection is a very precise tool, amplifying the effects of even the smallest phenotypic variations on reproductive success in ways that are always objectively rationalisable, sometimes generalisable and occasionally predictable. This 'paradox' may partly explain why some people find it difficult to understand the theory of evolution, despite its conceptual simplicity.

The Incompleteness of Biology

The twin pillars of modern biology, evolution and cell theory, have enabled us to construct a scientific account of the living world that can potentially explain the results of our experiments and field studies. In Kuhn's terminology, they constitute a paradigm within which all our theoretical and experimental activities are 'puzzle-solving'. However, they entail a cost. Modern biology includes no account of how evolution began or how the first cell was made.

In contrast to earlier 'transformists', Darwin explicitly excluded the origin of life from his theory, though he speculated briefly about it (Chapter 12).

Science deals with reproducible phenomena or patterns. It accounts for them by theories and it explores them by experiment or simulation. The origin of life on Earth was, in effect, a 'one-off' event; it is not covered by any theory in biology, physics or chemistry; and because the conditions under which it occurred are largely mysterious, it cannot be made the subject of experiment or simulation with any kind of reliability. So there are philosophical reasons for arguing that the origin of life is not a problem within the ambit of science. However, biologists are keenly aware of this profound lacuna in their view of the world. Therefore, the origin of life has attracted and continues to attract the attention of some of the best minds in the field. They can only speculate, but they do so from a position of detailed knowledge and deep understanding.

Unless and until these speculations generate a body of theory that leads to testable predictions and can be harmonised with the rest of our knowledge about the living world, biology will remain incomplete. But if a suitable theory ever arises from these endeavours, it will enable us not only to account for the origin of terrestrial life but also to make *rational* predictions about life elsewhere in the universe. That is not possible in our present state of understanding. The promise of so great an intellectual reward makes attempts at informed speculation about the origin of life worthwhile.

Biological Nature and Human Culture

Human institutions can no more be reduced to biology than can the process of construction of a wasp's nest be reduced to chemistry and physics. Popper[9] made a useful distinction between World 1 (the natural observable world), World 2 (mental representations of World 1) and World 3 (the products of World 2, such as poems and buildings and scientific theories). The contents of World 3, human culture, cannot be deduced or predicted from contents of World 1 such as human brain function. Although evolutionary theory has been evoked in discussions of general philosophical issues concerning ethics, the origins of religious traditions and the relationship of humans to the rest of the natural world, such issues are beyond the remit of biology. Evolutionary theory concerns World 1 and the issues of philosophy belong to World 3.

[9] Popper KR (1978) Three Worlds. The Tanner Lecture on Human Values, University of Michigan. http://www.tannerlectures.utah.edu/lectures/atoz.html#p (accessed 12 April 2008).

We have claimed in this book that science as a way of constructing knowledge is specifically a product of post-Renaissance western culture. Particular scientific ideas have arisen in particular socio-economic contexts (e.g. classical thermodynamics in the age of the steam engine), but the validity of an idea is independent of cultural circumstances. The *origin* of a belief is a matter of cultural history but its *validity* transcends cultural particulars. Science *as an institution* is the product of just one culture, but it is wrong (the 'genetic fallacy') to infer that particular theories, particular items of knowledge, have no better claim to credibility than the animistic beliefs of prehistoric people (see chapters 1 and 2).

'Cultural evolution' and biological evolution are fundamentally different, though both are unprogrammed and unpredictable. Cultural evolution is 'Lamarckian'. When a useful new characteristic is acquired – a new way of cooking or making clothes, a new weapon for war or hunting – it is transmitted to succeeding generations by teaching and learning, and to other social groups by (for instance) trading. Consequently, cultural transformation is far more rapid than biological transformation. But no long-term development in society was ever 'inevitable', though it might appear so in retrospect. Thus, the notion that scientific thinking is the inevitable concomitant of a sufficient level of social development is false. Human history refutes it. Only a very particular combination of social, political and economic circumstances can give rise to a way of thinking like science (Chapter 5), and those circumstances are not foreordained.

What happens when they cease to obtain? That question should concern us. Consider the modern developed world in the terms we applied to the ancient Greek city states (Chapter 2).

- Modern power and wealth are certainly in the hands of merchants, but not as they were in the ancient Greek cities or early capitalist Europe. Today's powerful merchants are not individuals or small groups but massive international companies, highly centralised and hierarchical.
- Increasing globalisation means that centres of trade have ceased to be cultural melting-pots in any dynamic, constructive way. In the modern developed world there is no imperative to assimilate radically new cultural beliefs, practices, attitudes or values, except those imposed by the sellers of new technologies.
- As the wealthy organisations – and the governments that seek to regulate them – become progressively larger, more centralised and more bureaucratic, freedom and independence of thought cease to be fostered. Bureaucracy is fatal to science.
- Competition among giant merchant groups rather than individuals fosters collective rather than individual ingenuity. It is surely no coincidence that the second half of the twentieth century saw the wholesale replacement of individual scientists, pursuing their own independent lines of research without time limits, by large organised teams working towards short-term goals driven by immediate commercial motives.
- Social communication as a whole is coming to depend less on the written word and increasingly on visual imagery, notably television and the internet. Visual imagery favours the concrete rather than the abstract.

It is a cliché that we live in a science-dominated world, but the foregoing points suggest that science as a way of thinking, as the phenomenally effective mode of knowledge production it has been for four centuries, may face extinction. Suppose it *were* to become extinct. Our culture's knowledge base, our set of beliefs about the world, would probably remain more or less intact. The validity of beliefs is independent of their roots. Our knowledge base could survive the loss of its roots just as, for example, the second law of thermodynamics survived the demise of the steam age. But how could that knowledge base continue to grow if its means of production had gone?

Living with Uncertainty

We can never predict the future of a species, or a culture, or an individual, but as humans we always try to plan ahead. Any society's approach to planning depends on its current beliefs and technology.[10] In the modern developed world, therefore, we feel that our future depends on science and its related technology. But that does not mean that science is universally loved and trusted. It often gives conflicting and uncertain answers to the world's problems (consider the current controversies about global warming or the safety of genetically modified crops) and the public reacts to such conflict and uncertainty with an amalgam of disappointment, consternation and scepticism. If science cannot provide definitive and unequivocal answers, what is the point of it? Since many people also believe that 'science' is responsible for world problems such as pollution, the exhaustion of resources and the destruction of habitat and species loss, mistrust is hardly surprising.

If the *nature* of science were better understood then its public image may achieve a better balance. Books introducing cosmology, quantum theory, evolutionary theory, cell biology and other aspects of science to the lay reader have enjoyed well-earned success, but few of them discuss what science *is*, how it originated, how it operates as a method for generating knowledge, what its limitations are. One of our aims in the present book has been to fill this gap.

Living with uncertainty is part of the human condition. Nowadays, in contrast to the rest of human history, the problems we face are largely global and threaten the future of our own and many other species world-wide; but thanks to the wealth of modern scientific knowledge and modern communications, we are more aware of them than we would have been in the past. Science affords our only means of addressing these problems rationally. To ask for simple, unequivocal and uncontentious solutions – or clear predictions – is to demand more than science is capable of providing, but it still represents our best hope.

---

[10] Yet there are important influences that we cannot foresee. Earthquakes, for example, have mostly been unpredictable and remain so. Will future knowledge provide us with new means of anticipating them? It has recently been suggested that some animals receive early warning of an impending earthquake by detecting the ultralow frequency emissions that occur at the start.

This makes the possibility that science will be extinguished as a mode of knowledge-production all the more worrying.

Human Beings as Moral Agents

Suppose we *could* attain coherent scientific accounts of one or more of the modern world's major problems. For example, suppose that in spite of all the practical difficulties we obtained such full and detailed knowledge of global warming that we could identify all its major causes and make quantitative predictions about what would happen if (a) our practices remained unchanged or (b) particular practices were altered or discontinued. What then? Making decisions about what to do and what not to do, and implementing those decisions, would remain moral and political issues, not scientific ones.

By and large, scientists are morally aware people. Most of them are deeply concerned about world problems such as global warming and often hold strong opinions about what should be done. Those opinions are informed but they do not follow from knowledge alone; they entail ethical judgments that are external to science.

There is nothing special about scientists in this regard. Given the same knowledge, most humans would form similar opinions. We are moral agents in a way that other species (even other species of apes) are not. The ability to anticipate and plan entails a capacity to judge some plans and their likely outcomes as better or more acceptable than others. That capacity has nothing to do with science *per se*.

One school of thought contradicts this contention, claiming that morality *does* fall within the remit of science and that 'ought' can be deduced from 'is'. The main proponents of this view are evolutionary psychologists, who believe that all aspects of human behaviour and society are, at root, ways of increasing reproductive success, i.e. are products of natural selection. Evolutionary psychologists seem to perceive no fundamental difference between cultural and biological evolution, and many of them embrace speculative explanations with low standards of proof. For example, they have sought to 'explain' crime as a male display of willingness to take risks, acne as a way of reducing attractiveness among those who are too young to care for children, and blushing as a female signal of readiness to mate – leaving open the question of why men blush. In none of these cases (or many others) is there any discernable evidence, which leads us to infer that evolutionary psychology, at least in these manifestations, is not science. It is trivially true that moral rules help to ensure the survival of the social group, and thus the transmission of one's genes to future generations. But that recognition does not *explain* moral rules, or account for the differences in such rules among cultures, or for changes in those rules. Also, it overlooks the fact that obedience to moral rules can sometimes *diminish* the probability of passing on one's genes. In our view, the tenuous nature of much of evolutionary psychology constitutes powerful evidence that ethics cannot be 'deduced from biology'.

Since humans are moral agents, our minds must choose among options. This raises an ancient philosophical problem: do we have free will or are our actions predetermined? Our minds *appear* to be free to make choices, but minds are 'caused' exclusively by brains, and brains are material objects that behave in physico-chemically determined ways. One possible resolution of this dilemma lies in the sheer complexity of the causal factors involved, which might make the brain-events involved in difficult moral choices 'chaotic'; i.e. the outcome is hypersensitive to minute changes in initial (causal) conditions. Thus, when we reflect on a moral choice, perhaps we marginally increase or decrease the activities in certain neural circuits, and that could suffice to change the outcome – but the outcome could not be predicted even in principle. 'Free will', in other words, might consist in the unpredictability of brain functions, which are wholly determined but may sometimes behave chaotically.

Biology, Humanity and World Problems

The human species is a product of evolution. You as an individual are the product of an unbroken sequence of cell divisions that has extended over some three thousand million years. In particular, the human mind is a product of evolution; it is a proper topic of study for biologists. That seems to trap us in a circular argument, a claim that the very entity that produces scientific explanations, the human mind, is amenable to scientific explanation. But the circularity is apparent rather than real. It confuses 'process' with 'description'. In Popperian terms, the evolutionary process that produced the human mind belongs to World 1, the description of that process to World 3.

In *About Life*, we claimed that the evolution of mind provides a scientific argument for the uniqueness of *Homo sapiens*. The emergence of the human mind, an entity capable of reflecting on and accounting for itself, added something truly novel and distinctive to the cosmos. If this most remarkable product of life on Earth is not capable of analysing and overcoming the problems that it faces, what else could be?

But our ability to analyse and overcome problems depends on our theories, and our present world problems do not fall within the domains of single or well-articulated bodies of theory. That, in essence, is the difficulty we face in planning for the future in our science-dominated, high-technology world. It will not be overcome easily.

Appendix: Science and Philosophy

Philosophies of Science and Scientific Practice

John Locke said that philosophy clears the undergrowth and permits science to grow. Yet many scientists are impatient of philosophy, even when they are lost in the undergrowth and armed with inadequate machettes. They remain impatient even when, in effect, they are doing philosophy themselves. The severance between philosophy and science has grown so deep that it is institutionalised: philosophy is seldom taught in university science faculties.

Many recent philosophers of science such as Morton Beckner and Dudley Schapere have recognised that overarching accounts of knowledge such as positivism and mechanistic materialism are not particularly useful because science is a heterogeneous enterprise. The disciplines we label 'science' today share a family resemblance but they involve different thinking styles and different ways of constructing knowledge, as we saw in Chapter 16. These philosophers have focused on the minutiae of actual scientific practice rather than 'grand theory'. They draw on long traditions in the philosophy of knowledge, but their work (Schapere's in particular) shows the influence of the history and sociology of science as well. Their investigations are pertinent to all scientists who wish to reflect on the nature of their own work.

The heterogeneity of science is manifest in inconsistencies. In Chapter 13 we described Morgan as inconsistent: he was a mechanistic materialist and an ardent experimentalist, but he believed in crossing-over when the hypothesis had no experimental support. Morgan is by no means alone. Scientists who adopt definite philosophical positions are seldom able to remain consistent.

This point is well illustrated by Einstein, who in 1905 published six papers of which at least three (the photoelectric effect, Brownian motion[1] and the special theory of relativity) were landmarks in the development of modern physics. In his autobiographical notes, Einstein stated that before writing those papers he had been influenced by the philosophical works of David Hume and Ernst Mach. If so, then he was sceptical about the Newtonian-Kantian account of the world, about induction, and about the knowability of 'things-in-themselves'.

---

[1] 'Brownian motion' is the rapid, random 'jumping' movements of microscopic particles suspended in a fluid such as water. The phenomenon was first observed in 1827 by the botanist Robert Brown (1773–1858) in small particles within pollen grains; hence the name.

P.S. Agutter, D.N. Wheatley, *Thinking about Life*,
© Springer Science+Business Media B.V. 2008

Mach dismissed the notions of absolute space and absolute time because they have no basis in the evidence of the senses. Einstein's special theory of relativity paper accounted for the non-Newtonian behaviour of objects moving at very high velocities by accepting Mach's position. The new theory replaced Newtonian mechanics in certain applications.[2] But for the same philosophical reasons, Mach also dismissed belief in atoms; such was his influence that leading chemists of the day such as Oswald also rejected atomic theory. If Einstein had been philosophically consistent, he would have been just as sceptical about atoms as he was about the Newtonian-Kantian idea of absolute space and time. But he was not.

In his paper on Brownian motion, Einstein proved that atomic theory provides the *only* perspective capable of accounting for that phenomenon. The paper ended with an invitation, or challenge, to experimentalists to test his mathematical predictions. Two years later, Perrin did so; atomic theory was firmly established. Einstein had effectively proved the existence of atoms, the 'metaphysical' concept that Mach and his followers had scorned.[3]

These examples show that for the most eminent of scientists, no single philosophical account of 'science' suffices for all occasions. General philosophical positions can be valuable guides but they should not become straitjackets. We do not blame Einstein for inconsistency. Neither, therefore, should we blame Morgan, or any other scientist in a similar situation.

The Nature of Scientific Theories

Both the aforementioned Einstein papers challenged entrenched beliefs (Newtonian mechanics in one case, and the invalidity of atomic theory on the other). But how *deeply* was each of these beliefs entrenched? According to Kant, a concept is undeniable if it is so fundamental to thought that it seems innate in the mind's structure. Therefore, to challenge a *deeply* entrenched concept is to oppose Kant, i.e. to share the scepticism of Hume and Mach. Hence the extreme phenomenalism of the special relativity paper. In contrast, although atomic theory is a deep issue, it

---

[2] Einstein observed (apparently at the age of 16) that Maxwell's equations would give infinite solutions if the observer was travelling at the same velocity as a light wave. That was tantamount to saying that the equations did not apply at that velocity, i.e. no observer could travel at the speed of light. This was the main motivation for the special theory of relativity. The paper was directly relevant to the physics of the time; it explained the unexpected result of the Michaelson-Morley experiment (which effectively disproved the existence of the 'luminiferous aether') and justified the Lorenz-Fitzgerald contraction, an *ad hoc* device for explaining the 'failure' of the Michaelson-Morley experiment.

[3] Strictly speaking, Einstein had validated Boltzmann's *kinetic theory*, but as phenomenalists such as Mach had pointed out, kinetic theory made no sense if atoms did not exist. In 1908, a year after Perrin's work was published, the phenomenalist Wilhelm Ostwald – previously an implacable opponent of atomism – publicly declared his conversion.

is not 'part of the mind's innate structure'. Therefore, the Brownian motion paper was compatible with Kant and failed to show the influence of Hume and Mach. This contrast between 'deep' and 'less deep' ideas becomes clearer when we consider how theories are constructed and how they change.

The word 'theory' can be problematic. For one thing, it is not used consistently even by philosophers or scientists. The most satisfactory brief definition we can offer is: *a set of data, general principles and postulates interlinked by logical or mathematical argument and affording a general means for explaining and predicting a wide range of phenomena.* All major theories conform to this broad definition: Newtonian mechanics, thermodynamics, electromagnetic theory, relativity, quantum mechanics – and evolutionary theory. They are logically coherent conceptual structures founded upon postulates and definitions. They consist of laws of nature, procedures of reasoning and relevant data. They interconnect indefinitely large ranges of observations and experimental results that could otherwise seem unrelated, such as the falling of apples and the orbiting of planets. They are all *dynamic* and continually changing.

Generalisations and principles are fairly uncontentious aspects of theory, but why do we include *data*? The reason is that scientific data are not just *any* data; they are selected. When an object is described in Newtonian mechanics, the relevant properties include mass, position, initial velocity and acceleration. We ignore other properties, which might be important to us in different contexts (e.g. colour, odour, monetary value, whether it is a common or unusual object). *A theory obliges us to consider only certain data and to exclude others.* In Newtonian mechanics, mass and velocity are independent of one another; in relativistic mechanics they are not, and we must speak of the 'rest mass' of an object (defined as the mass when its velocity relative to the observer is zero). Thus, what data are relevant, and how they are defined and interpreted, depend on the theory. *Data must be regarded as parts of theory.*

That may seem contrary to intuition, even to common sense. Let us consider an everyday observation statement. Surely 'the dog is asleep on the sofa' remains true or false no matter what theories we accept? (We might have different opinions about whether the dog *ought* to be asleep on the sofa, but that is another matter.) However, 'the dog is asleep on the sofa' is unproblematic only because we all agree about the use of English words and syntax. If we did not, then two people might argue about whether the statement was true, or whether it meant anything at all.

In Chapter 1 of this book we noted that although 'science' is in some respects a special kind of knowledge, it involves the same perceptions, dispositions and mental processes that we use in everyday life. Thus, we may regard our ordinary native language as representing our 'theory' of the everyday world. If we use a different language we have a different 'theory of the world'. Analogously, we may regard a scientific theory as a specialised language for talking about particular *aspects* of the world: different theories are expressed in different 'languages'. Alternative theories such as Newtonian and relativistic mechanics may make different sense of apparently identical observations.

The psychologist George Kelly epitomised this position in his phrase 'man as scientist'. According to Kelly,[4] we start to form mental representations of the world around us during infancy and we continually test those representations against experience and modify them as required. The language in which we talk about daily experience, our own native language, is a dynamic part of that mental representation: it shapes it and is shaped by it. In other words, it behaves just as a scientific theory behaves.

What we observe and recognise depends on the way our minds construe the world, not just on the images conveyed to our brains from our sense-organs. To that extent, we have to agree with Kant. For instance, people from Western culture can interpret a picture showing an array of parallel lines as a staircase (seen either from above or from below). But people from many African cultures cannot interpret the image that way. They do not construe pictorial representations as post-Renaissance Europeans do. Our language and our cultural habits enable us to understand the world but they also *constrain* our understanding. The theories used by scientists are their languages and cultural habits. Theories enable us to understand the part of the world we are studying, but they also constrain our understanding.

Do these considerations help us to make sense of the claim that theoretical ideas in science may be 'deep' or 'less deep'?

Theory Structure and Theory Change[5]

We can picture a theory as an amoeba that lives in a sea of data and grows by engulfing and assimilating all the facts that are compatible with its 'metabolism' (its postulates, definitions, laws and procedures of reasoning). Amoebae continually move around, changing their shapes and growing, unless they are dead. Scientific theories are ever-shifting and ever-growing, unless they are dead. Amoebae examine and ingest possible morsels of food by extending pseudopodia. Theories examine and ingest possible morsels of data by extending hypotheses. At its periphery, the structure is very labile. Parts of the surface can be removed and the amoeba (or the theory) will re-seal and carry on as before, more or less unaffected. At the core, around the 'nucleus', things are different. Removal or alteration of material here has dramatic effects. The theory is radically changed or killed. This metaphor gives a visual impression of 'deep' and 'less deep' theoretical ideas.

Bigger and more active amoebae sometimes eat smaller ones. Analogously, two theories might 'fuse' to produce a more general one. Newtonian mechanics ingested

---

[4] The book by Fransella and Bannister (see bibliography) gives a lucid introduction to Kelly's theory of personality.

[5] The 'amoeba' metaphor in this section is based on Willard Quine's description of scientific theories. The 'natural selection' analogue of the evolution of theories is due to Stephen Toulmin. See the bibliography for references.

and assimilated Kepler's account of the solar system. Classical thermodynamics and classical mechanics were combined in statistical mechanics. If two theories or two variants of a theory co-exist they may compete, whereupon the better-adapted one survives. 'Better-adapted' means better able to assimilate and rationalise new data and better equipped to generate testable hypotheses. In the middle of the 20th century, two rival theories of cosmology were current; they came to be known as the 'big bang' and 'steady state' theories. During the third quarter of the century, more and more new data, such as the constant background microwave radiation of the universe, proved incompatible with the steady state theory and consistent with the predictions of the 'big bang'. As a result, the 'big bang' came to be more and more widely accepted while the steady state theory was gradually abandoned. That is not to say that the 'big bang' theory is *true* in any absolute sense; rather, the theory is *useful* in a way that the steady state theory is not.

Theories can no more be adjudged 'true' or 'false' than languages can. But a theory, like a language, might prove more or less useful, and it is the useful ones that survive. We accept, deploy and believe *useful* theories. Thus, we cannot claim that Einstein's theory is true and Newton's false; rather, Einstein's theory is useful in contexts where Newton's theory is not. Nor, more subtly, is Newton's theory a 'special case' of Einstein's, i.e. Einstein's theory reduces to it when certain restrictions are applied. Its foundations and procedural rules are quite different. These theories continue to co-exist because they both remain useful for different purposes.

Experiments

New data, and tests of new hypotheses, depend on observation and experiment. An experiment is a way of making precise observations, not of 'unconstrained nature' but of a situation deliberately constrained so that all relevant variables are known. The ability to design and conduct experiments is as essential for a scientist as a thorough and up-to-date understanding of relevant theories.

A properly designed experiment must be *reproducible*. That is to say, it must give substantially the same results when it is repeated at a different time and place and by other (trained and competent) experimenters. It must also be *valid*: anyone trained in the appropriate field of science must agree that it does just what its designer claims it does. If new techniques or equipment are involved then the experimenter must show that these meet appropriate standards of reliability, independent of times, places and persons. The results of the experiment must be *interpretable*: they must be free of 'interfering variables' and of errors of extrapolation or interpolation. This can be ensured by running *controls*, in which all variables except the one under investigation are kept at the same values as in the experiment itself. It is also important to ensure that a newly-designed experiment is practicable, ethical and economic.

Good experiment design is an art that requires practice. As with drawing or creative writing the basic principles can be taught (though they are not always taught well), but real skill is achieved only after years of effort and of learning from mistakes. Also – again, as with drawing and creative writing – the capacity for clear and careful observation and for constructive self-criticism are fundamental to the acquisition of technique.

The principles of experiment design and experimental practice have seldom engaged the attention of philosophers to the extent that issues of theory, hypothesis, truth and logic have. This may be an area on which more philosophers of science could profitably focus.

Models

Scientists and philosophers often use the word 'model' in reference to matters of both theory and experiment. 'Model', like 'theory', has a number of meanings in English, most of which are irrelevant to science. For instance, it can denote: a person who poses for a photograph or painting; a preparatory sculpture to aid the design for a finished work; an original unique article of clothing; design or style; or a standard to be emulated. In science, 'model' may have one of the following meanings:

- A (small-scale) representation of structure or device intended to illustrate or test the properties of the real thing. It is sometimes impractical or uneconomical to perform experiments or even to make reliable observations on 'the real thing', so the investigators simulate it, for example by computer *models*.
- An idealised representation of an object or situation. The 'frictionless surface' of Newtonian mechanics and the 'ideal gas' of classical thermodynamics do not exist in the world of 'things-in themselves'. They are fictions to which the equations of the relevant theories apply exactly. Their role is to tell us how the real world would behave if it were less imperfect; the real world becomes comprehensible in terms of (usually minor) deviations from such ideals.
- A simplified representation or description of a complex entity. All details of the real entity are deleted from the description unless they are essential for making sense of its behaviour. An organic chemist, for example, might model a reaction in terms only of the functional groups involved, no matter how complex the rest of the molecule. Functional groups are almost always simple arrangements of few atoms. This practice simplifies the description of the reaction and emphasises similarities with other reactions of the same general type. The pattern might be impossible to discern if *all* the details of the reacting molecules were included.

The developmental psychologist Jerome Bruner wrote interestingly on the subject of models. He pointed out that models may be *symbolic, iconic* or *enactive*. In science, a 'symbolic' model is manifest in words, mathematical formulae or other abstractions, an 'iconic' model in pictures. A gene may be represented as a string of letters, each denoting a nucleotide (symbolic model); the DNA double helix is a familiar visual image (iconic model). The main purpose of 'enactive models' is to teach skills; you cannot effectively teach a child to ride a bicycle by words or pictures. In science, enactive models have a key role in teaching students how to design and execute experiments, or to use specialised equipment.

Bibliography

Chapter 1

Beck WS (1957) Modern Science and the Nature of Life. Harcourt, Brace, New York.
Bronowski J (1978) The Common Sense of Science. Harvard University Press, Cambridge, MA.
Haraway D (1989) Primate Visions. Routledge, London.
Medawar PB (1982) Pluto's Republic. Oxford University Press, New York.
Medawar PB (1985) The Limits of Science. Oxford University Press, Oxford.
Marsak LM (ed) (1964) The Rise of Science in Relation to Society. Macmillan, New York.
Menard HW (1971) Science: Growth and Change. Harvard University Press, Cambridge, MA.
Moore JA (1993) Science as a Way of Knowing: The Foundations of Modern Biology. Harvard University Press, Cambridge, MA.
Santayana G (1962) Reason in Science. Collier, New York.
Turchin VE (1977) The Phenomenon of Science. University of Columbia Press, New York.
Ziman J (1978) Reliable Knowledge. Cambridge University Press, Cambridge.

Chapter 2

Barnes B (1974) Scientific Knowledge and Sociological Theory. Routledge & Kegan Paul, London.
Barnes B, Shapin S (eds) (1979) Natural Order: Historical Studies of Scientific Culture. Sage, London.
Basalla G (1988) The Evolution of Technology. Cambridge University Press, Cambridge.
Ben-David J, Clark TN (eds) (1977) Culture and Its Creators. University of Chicago Press, Chicago, IL.
Bloor D (1976) Knowledge and Social Imagery. Routledge & Kegan Paul, London.
Cole M, Scribner S (1974) Culture and Thought. Wiley, New York.
Dunbar R, Knight C, Power C (eds) (1999) The Evolution of Culture. Edinburgh University Press, Edinburgh.
Fehl NE (1965) Science and Culture. Chu Ching, Hong Kong.
Forbes RJ (1955–64) Studies in Ancient Technology (9 volumes). Brill, Leiden.
Gellner E (1974) Legitimation of Belief. Cambridge University Press, Cambridge.
Hawkes J (1973) The First Great Civilizations: Life in Mesopotamia, the Indus Valley, and Egypt. Knopf, New York.
Jarvie IC (1972) Concepts and Society. Routledge & Kegan Paul, London.
Mulkay M (1985) The Word and the World. Allen & Unwin, London.

Resnikoff HL, Wells RO (1973) Mathematics and Civilization. Holt, Reinhart & Winston, New York.
Rose HA, Rose SPR (1976) The Political Economy of Science. Macmillan, London.

Chapter 3

Canfora L (trans. Ryle M) (1989) The Vanished Library. A Wonder of the Ancient World. University of California Press, Berkeley, CA.
Hare RM (1982) The Philosophy of Plato. Oxford University Press, Oxford.
Lloyd GER (1970) Early Greek Science; Thales to Aristotle. Norton, New York.
Lloyd GER (1991) Methods and Problems in Greek Science. Cambridge University Press, Cambridge.
Macleod R (ed) (2004) The Library of Alexandria: Centre of Learning in the Ancient World. IB Tauris, London.
Ross WD (1959) Aristotle: A Complete Exposition of His Works and Thought. Meridian, Cleveland, OH.
Stahl W (1962) Roman Science. University of Wisconsin Press, Madison, WI.
Waterlow S (1982) Nature, Change and Agency in Aristotle's Physics. Clarendon Press, Oxford.

Chapter 4

Arnold T, Guillaume A (1931) The Legacy of Islam. Oxford University Press, Oxford.
Ajram K (1992) The Miracle of Islamic Science. Cedar Graphics, Cedar Rapids, Iowa, IA.
Bakar O (1999) History and Philosophy of Islamic Science. Islamic Texts, Cambridge.
Baldwin JW (1971) The Scholastic Culture of the Middle Ages, 1000–1300. Heath and Co., Lexington, MA.
Benson RL, Constable G (1991) Renaissance and Renewal in the Twelfth Century. University of Toronto Press, Toronto.
Coplestone FC (1955) Aquinas. Penguin, Baltimore, MD.
Crombie AC (1971) Robert Grosseteste and the Origins of Experimental Science 1100–1700. Clarendon Press, Oxford.
Grant E (1996) The Foundations of Modern Science in the Middle Ages: Their Religious, Institutional and Intellectual Contexts. Cambridge University Press, Cambridge.
Huff T (1995) The Rise of Early Modern Science: Islam, China and the West. Cambridge University Press, Cambridge.
Lindberg DC (1992) The Beginnings of Western Science. University of Chicago Press, Chicago, IL.
Nasr SN (1968) Science and Civilization in Islam. New American Library, New York.
Pines S (1986) Studies in Arabic Versions of Greek Texts and in Mediaeval Science. Brill, Leiden.
Riché P (1976) Education and Culture in the Barbarian West: From the Sixth through the Eighth Century. University of South Carolina Press, Columbia.
Russell B (1946) History of Western Philosophy, 3rd ed. George Allen & Unwin, London.
Sarton G (1927) Introduction to the History of Science. Carnegie Institution of Washington Publication no. 376, Baltimore, MD.
Turner HR (1995) Science in Medieval Islam: An Illustrated Introduction. University of Texas Press, Austin, TX.
White TH (1954) The Bestiary: A Book of Beasts, being a Translation from a Latin Bestiary of the Twelfth Century. Putnam's, New York.

Chapter 5

Armitage A (1957) Copernicus, the Founder of Modern Astronomy. Thomas Yoseloff, New York.
Armitage A (1966) John Kepler. Faber, London.
Buchwald JZ, Bernard I (eds) (2001) Isaac Newton's Natural Philosophy. MIT Press, Cambridge, MA.
Dannenfeldt KH (ed) (1974) The Renaissance: Basic Interpretations. Heath and Co., Lexington, MA.
Drake S (1957) Discoveries and Opinions of Galileo. Doubleday, New York.
Drake S, Drabkin IE (1969) Mechanics in Sixteenth-Century Italy. University of Wisconsin Press, Madison, WI.
Dreyer JLE (1890) Tycho Brahé: A Picture of Scientific Life and Work in the Sixteenth Century. Black, Edinburgh.
Englander D, Norman D, O'Day R, Owens WR (eds) (1979) Culture and Belief in Europe 1450–1600. Blackwell, Oxford.
Hall AR (1983) The Revolution in Science 1500–1750. Longman, London.
Hall AR (1996) Isaac Newton: Adventurer in Thought. Cambridge University Press, Cambridge.
Hallyn F (1990) The Poetic Structure of the World: Copernicus and Kepler. Zone Books, New York.
Henry J (2002) Knowledge is Power: Francis Bacon and the Method of Science. Icon Books, Cambridge.
Kenny A (1968) Descartes: A Study of His Philosophy. Random House, New York.
Lattis JM (1995) Between Copernicus and Galileo. University of Chicago Press, London.
Lindsay D, Price MR (1975) Authority and Challenge: A Portrait of Europe 1300–1600. Oxford University Press, London.
McKnight SA (ed) (1992) Science, Pseudo-Science and Utopianism in Early Modern Thought. University of Missouri Press, Columbia, MO.
Murray A (1978) Reason and Society in the Middle Ages. Clarendon, Oxford.
Pitt JC (1997) Galileo, Human Knowledge and the Book of Nature: Method Replaces Metaphysics. Kluwer, Dordrecht.
Porter R, Teich M (eds) (1992) The Scientific Revolution in National Context. Cambridge University Press, Cambridge.
Rothenstein J (1964) Francis Bacon: An Introduction. Thames and Hudson, London.
Runciman S (1965) The Fall of Constantinople. Cambridge University Press, London.
Shapere D (1974) Galileo: A Philosophical Study. University of Chicago Press, Chicago, IL/London.
Shea WR (1972) Galileo's Intellectual Revolution. Macmillan, London.
Thijssen JMMH, Zupko J (eds) (2001) The Metaphysics and Natural Philosophy of John Buridan. Brill, Leiden.
Urbach P (1986) Francis Bacon's Philosophy of Science: An Account and a Reappraisal. Open Court, La Salle, IL.
Waley D (1964) Later Mediaeval Europe. Longmans, Green and Co., London.
Wallace WA (1991) Galileo, the Jesuits and the Mediaeval Aristotle. Ashgate, Aldershot.
Westfall RS (1980) Never at Rest: A Biography of Isaac Newton. Cambridge University Press, Cambridge.
Wightman WPD (1972) Science in a Renaissance Society. Hutchinson, London.
Williams B (1990) Descartes: The Project of Pure Enquiry. Penguin, London.
Yates FA (1964) Giordano Bruno and the Hermetic Tradition. Routledge & Kegan Paul, London.

Chapter 6

Bradbury S (1967) The Evolution of the Microscope. Pergamon, Oxford/London.
Cushing H (1943) A Bio-Bibliography of Andreas Vesalius. Schuman's, New York.

Hooke R (1665) Micrographia. Royal Society, London.

Espinasse M (1962) Robert Hooke. University of California Press, Berkeley, CA.

Fuchs T (trans. Grene MG) (2002) The Mechanization of the Heart: Harvey and Descartes. Boydell and Brewer, Rochester, NY.

Graubard M (1964) Circulation and Respiration: The Evolution of an Idea. Harcourt, Brace & World, New York.

Gregory A (2001) Harvey's Heart, The Discovery of Blood Circulation. Icon Books, Cambridge.

Kuhn TS (1970) The Structure of Scientific Revolutions, 2nd ed. University of Chicago Press, Chicago, IL.

Pagel W (1983) New Light on William Harvey. Transaction Publishers, Pitscataway, NJ.

Rapson H (1982) The Circulation of the Blood. Frederick Muller, London.

Scott JF (1976) The Scientific Writings of René Descartes. Taylor & Francis, London.

Siegel RE (1968) Galen's Systems of Physiology and Medicine. S. Karger, Basel.

Wilson C (1995) The Invisible World: Early Modern Philosophy and the Invention of the Microscope. Princeton University Press, Princeton, NJ.

Chapter 7

Barnes J (ed) (1995) Cambridge Companion to Aristotle. Cambridge University Press, Cambridge.

Downey G (1962) Aristotle: Dean of Early Science. Franklin Watts, New York.

Gotthelf A, Lennox J (eds) (1987) Philosophical Issues in Aristotle's Biology. Cambridge University Press, Cambridge.

Gotthelf A (ed) (1985) Aristotle on Nature and Living Things. Philosophical and Historical Studies. Mathesis, Pittsburgh, PA.

Pellegrin P (1986) Aristotle's Classification of Animals. Biology and the Conceptual Unity of the Aristotelian Corpus. University of California Press, Berkeley, CA.

Preus A (1975) Science and Philosophy in Aristotle's Biological Works. Georg Olms, New York.

Chapter 8

Fellows OE, Milliken SF (1972) Buffon. Twayne, New York.

Foster M (1901) Lectures on the History of Physiology during the Sixteenth, Seventeenth and Eighteenth Centuries. Cambridge University Press, Cambridge.

Gloyne SR (1950) John Hunter. Livingstone, London.

King LS (1963) Growth of Medical Thought. University of Chicago Press, Chicago, IL.

King LS (1970) The Road to Medical Enlightenment 1650–1695. Macdonald, London.

Leicester HM, Klickstein HS (1952) A Source Book in Chemistry 1400–1900. McGraw-Hill, New York.

Lindeborn GA (1968) Herman Boerhaave: The Man and his Work. Methuen, London.

Pagel W (1958) Paracelsus: An Introduction to Philosophical Medicine in the Era of the Renaissance. S. Karger, Basel.

Partington JR (1957) A Short History of Chemistry, 3rd ed. Macmillan, London.

Spellman WM (1997) John Locke. Macmillan, London.

Terrall M (2002) The Man Who Flattened the Earth: Maupertuis and the Sciences in the Enlightenment. University of Chicago Press, Chicago, IL.

Vartanian A (1960) La Mettrie's 'L'Homme Machine': A Study in the Origins of an Idea. Princeton University Press, Princeton, NJ.

Wheeler LR (1939) Vitalism: Its History and Validity. Witherby, London.
Wolf A (1952) A History of Science, Technology and Philosophy in the Eighteenth Century, 2nd ed. Allen & Unwin, London.
Woolhouse RS (1983) Locke. University of Minnesota Press, Minneapolis, MN.

Chapter 9

Appel T (1987) The Cuvier-Geoffroy Debate: French Biology in the Decades Before Darwin. Oxford University Press, Oxford.
Baker JR (ed) (1988) The Cell Theory: A Restatement, History and Critique. Garland, New York.
Bennett J (1971) Locke, Berkeley, Hume: Central Themes. Clarendon, Oxford.
Berlin I (ed) (1956) The Age of Enlightenment. Books for Libraries Press, Freeport, NY.
Blackmore JT, Itagaki R, Tanaka S (2001) Ernst Mach's Vienna 1895–1930: Or Phenomenalism as Philosophy of Science. Kluwer, Dordrecht.
Bowie A (1993) Schelling and Modern European Philosophy: An Introduction. Routledge, London.
Guyer P (1987) Kant and the Claims of Knowledge. Cambridge University Press, New York.
Eccles JC (1979) Sherrington, His Life and Thought. Springer, Berlin, MA/New York.
Ellington JW (1985) Kant's Philosophy of Material Nature. Hackett, Indianapolis, IN.
Faber RJ (1987) Clockwork Garden: On the Mechanistic Reduction of Living Things. University of Massachusetts Press, London.
Guyader H (trans. Grene MG) (2004) Étienne Geoffroy Saint-Hilaire, 1772–1844: A Visionary Naturalist. University of Chicago Press, Chicago, IL.
Harris H (1999) The Birth of the Cell. Yale University Press, London.
Holmes FC (1974) Claude Bernard and Animal Chemistry. Harvard University Press, Cambridge, MA.
Hunemann P (ed) (2007) Understanding Purpose: Kant and the Philosophy of Biology. University of Rochester Press, Rochester, NY.
Hyland P (2003) The Enlightenment. Routledge, London.
Kemp J (1968) The Philosophy of Kant. Oxford University Press, New York.
Knight DM (1998) Science in the Romantic Era. Ashgate, Aldershot.
Letwin SR (1965) The Pursuit of Certainty: David Hume, Jeremy Bentham, John Stuart Mill. Cambridge University Press, Cambridge.
Popper KR (1959) The Logic of Scientific Discovery. Hutchinson, London.
Richards R (2002) The Romantic Conception of Life: Science and Philosophy in the Age of Goethe. University of Chicago Press, Chicago, IL.
Rothschuh KR (1971) Du Bois-Reymond. In: Gillespie CC (ed) Dictionary of Scientific Biography, vol. IV, pp. 200–205. American Council of Learned Societies, Charles Schribner's Sons, New York.
Sigerist HE (1932) The Great Doctors: A Biographical History of Medicine. Books for Libraries Press, Freeport, NY. (Contains a short biography of Müller.)
Turner RS (1972) Helmholtz. In: Gillespie CC (ed) Dictionary of Scientific Biography, vol. VI, pp. 241–253. American Council of Learned Societies, Charles Schribner's Sons, New York.
Viëtor K (1950) Goethe the Thinker. Harvard University Press, Cambridge, MA.
Wright JP (1983) The Sceptical Realism of David Hume. Manchester University Press, Manchester.

Chapter 10

Child CM (1941) Patterns and Problems of Development. University of Chicago Press, Chicago, IL.
Davidson EH (1986) Gene Activity in Early Development, 3rd ed. Academic, New York.
De Pomerai D (1985) From Gene to Animal: An Introduction to the Molecular Biology of Animal Development. Cambridge University Press, New York.

Gasking EB (1967) Investigations into Generation. 1651–1828. History of Scientific Ideas. Hutchinson, London.

Gilbert SF (2006) DevBio: A Companion to Developmental Biology, 8th ed. Sinauer, Sunderland, MA.

Gould SJ (1977) Ontogeny and Phylogeny. Harvard University Press, Cambridge, MA.

Horder TJ, Witkowski JA, Wyle CC (eds) (1986) A History of Embryology. Cambridge University Press, Cambridge.

Kumé M, Katsuma D (1988) Invertebrate Embryology. Garland, New York.

Oppenheimer JM (1963) Essays in the History of Embryology and Biology. MIT Press, Cambridge, MA.

Roe SA (1981) Matter, Life and Generation: 18th Century Embryology and the Haller-Wolff Debate. Cambridge University Press, Cambridge.

Chapter 11

Belloni L (1975) Redi. In: Gillespie CC (ed) Dictionary of Scientific Biography, vol. XI, pp. 341–343. American Council of Learned Societies, Charles Schribner's Sons, New York.

Bulloch W (1960) The History of Bacteriology, 2nd ed. Oxford University Press, London.

Dobell C (1932) Anthony van Leeuwenhoek and His Little Animals. Bale and Danielsson, London.

Farley J (1977) The Spontaneous Generation Controversy from Descartes to Oparin. Johns Hopkins University Press, Baltimore, MD.

Gasking E (1967) Investigations into Generation 1651–1828. Johns Hopkins University Press, Baltimore, MD.

Harris H (2002) Things Come to Life. Oxford University Press, Oxford.

Chapter 12

Bowlby J (1990) Charles Darwin: A New Life. Norton, New York.

Bowler P (1974) Fossils and Progress: Paleontology and the Idea of Progressive Evolution in the Nineteenth Century. Science History, New York.

Bowler P (2003) Evolution: The History of an Idea, 3rd ed. University of California Press, Baltimore, MD.

Burchfield J (1975) Lord Kelvin and the Age of the Earth. University of Chicago Press, Chicago, IL.

Burkhardt R (1977) The Spirit of System: Lamarck and Evolutionary Biology. Harvard University Press, Cambridge, MA.

Cloyd EL (1972) James Burnett, Lord Monboddo. Clarendon, Oxford.

Corsi P, trans. Mandelbaum J (1988) The Age of Lamarck: Evolutionary Theories in France, 1790–1830.: University of California Press, Berkeley, CA.

Dalrymple GB (1991) The Age of the Earth. Stanford University Press, Stanford, CA.

Dean DR (1992) James Hutton and the History of Geology. Cornell University Press, Ithaca, NY.

Desmond A, Moore J (1991) Darwin. Michael Joseph, London.

Gillespie CC (1959) Genesis and Geology: The Impact of Scientific Discoveries upon Religious Beliefs in the Decades before Darwin. Harper & Row, New York.

Hodge MJS, Radick G (eds) (2003) The Cambridge Companion to Darwin. Cambridge University Press, Cambridge.

Hull D (ed) (1973) Darwin and His Critics: The Reception of Darwin's Theory of Evolution by the Scientific Community. University of Chicago Press, Chicago, IL.

Jordanova LJ (1964) Lamarck. Oxford University Press, London.

King-Hele D (1968) The Essential Writings of Erasmus Darwin. MacGibbon and Kee, London.

Ospovat D (1981) The Development of Darwin's Theory: Natural History, Natural Theology, and Natural Selection. Cambridge University Press, Cambridge.

Ryan A (1974) J. S. Mill. Routledge & Kegan Paul, London.

Shanahan T (2004) The Evolution of Darwinism: Selection, Adaptation, and Progress in Evolutionary Biology. Cambridge University Press, Cambridge.

Sheppard T (1920) William Smith. Brown & Sons, Hull.

Young RM (1985) Darwin's Metaphor: Nature's Place in Victorian Culture. Cambridge University Press, Cambridge.

Chapter 13

Allen GE (1978) Thomas Hunt Morgan: The Man and His Science. Princeton University Press, Princeton, NJ.

Bowler PJ (1983) The Eclipse of Darwinism: Anti-Darwinian Evolution Theories in the Decades Around 1900. Johns Hopkins University Press, Baltimore, MD.

Bowler PJ (1989) The Mendelian Revolution. Athlone, London.

Darden L (1991) Theory Change in Science: Strategies from Mendelian Genetics. Oxford University Press, New York.

Dunn LC (1965) A Short History of Genetics: The Development of some of the Main Lines of Thought, 1864–1939. McGraw-Hill, New York.

Jacob F (1973) The Logic of Life: A History of Heredity. Pantheon, New York.

Mackenzie M (1981) Statistics in Britain, 1866–1930. Edinburgh University Press, Edinburgh.

Moore J (1981) The Post Darwinian Controversies. Cambridge University Press, Cambridge.

Nyhart L (1995) Biology Takes Form: Animal Morphology and the German Universities 1800–1900. University of Chicago Press, Chicago, IL.

Olby RC (1966) The Origins of Mendelism, 2nd ed. Schocken Books, New York.

Sapp J (1987) Beyond the Gene: Cytoplasmic Inheritance and the Struggle for Authority in Genetics. Oxford University Press, New York.

Stern C, Sherwood ER (eds) (1966) The Origins of Genetics. A Mendel Source Book. W. H. Freeman, San Francisco, CA/London.

Sturtevant AH (1965) A History of Genetics. Harper, London.

Vitezslav O, trans. Flynn S (1996) Gregor Mendel: The First Geneticist. Oxford University Press, New York.

Chapter 14

Bendall DS (ed) (1983) Evolution from Molecules to Men. Cambridge University Press, Cambridge.

Dobzhansky T, Ayala F, Stebbins GL, Valentine JW (1977) Evolution. W. H. Freeman, San Francisco, CA.

Ghiselin MT (1969) The Triumph of the Darwinian Method. University of California Press, Berkeley, CA.

Glick T, Kohn D (eds) (1996) On Evolution: The Development of the Theory of Natural Selection. Hackett, Indianapolis, IN.

Kimura M (1983) The Neutral Theory of Molecular Evolution. Cambridge University Press, New York.

Laubichler M, Maienschein J (eds) (2007) From Embryology to Evo-Devo: A History of Developmental Evolution. MIT Press, Boston, MA.

Mayr E (1991) One Long Argument: Charles Darwin and the Genesis of Modern Evolutionary
    Thought. Harvard University Press, Cambridge, MA.
Mayr E, Provine W (eds) (1998) The Evolutionary Synthesis: Perspectives on the Unification of
    Biology, revised ed. Harvard University Press, Cambridge, MA.
Moore JA (1972) Heredity and Development, 2nd ed. Oxford University Press, New York.
Provine WB (1971) The Origins of Theoretical Population Genetics. University of Chicago Press,
    Chicago, IL.
Provine WB (1986) Sewell Wright and Evolutionary Biology. University of Chicago Press,
    Chicago, IL.
Smocovitis VB (1996) The Evolutionary Synthesis and Evolutionary Biology. Princeton
    University Press, Princeton, NJ.
Sober E (1984) The Nature of Selection: Evolutionary Theory in a Philosophical Focus. MIT
    Press, Cambridge, MA.

Chapter 15

Allen C, Lauder G (eds) (1993) Nature's Purposes. MIT Press, Cambridge, MA.
Crick F (2004) Of Molecules and Men. Promethius Books, New York.
Mayr E (1972) The Growth of Biological Thought: Diversity, Evolution and Inheritance. Harvard
    University Press, Cambridge, MA.
Monod J (1970) Chance and Necessity. Random House, New York.
Nagel E (1979) Teleology Revisited, and Other Essays in the Philosophy and History of Science.
    Columbia University Press, New York.
Taylor R (1966) Action and Purpose. Prentice-Hall, Englewood Cliffs, NJ.
Woodfield A (1976) Teleology. Cambridge University Press, Cambridge.
Wright L (1976) Teleological Explanation. University of California Press, Berkeley, CA.

Chapter 16

Beckner M (1959) The Biological Way of Thought. Columbia University Press, New York.
Bonner JT (1988) The Evolution of Complexity by Means of Natural Selection. Princeton
    University Press, Princeton, NJ.
Einstein A, Infeld L (1938) The Evolution of Physics. Simon & Schuster, New York.
Farber P (1994) The Temptations of Evolutionary Ethics. University of California Press, Berkeley, CA.
Gould SJ (2002) The Structure of Evolutionary Theory. Harvard University Press, Cambridge, MA.
Hull D (1998) (ed) Ruse M. Philosophy of Biology. Oxford University Press, Oxford.
Jones S (1999) Almost Like a Whale. Doubleday, London.
Maienschein J, Ruse M (eds) (1999) Biology and the Foundations of Ethics. Cambridge University
    Press, Cambridge.
Ruse M (1973) The Philosophy of Biology. Hutchinson, London.
Sober E (1993) Philosophy of Biology. Westview Press, Boulder, CO/San Francisco, CA.

Appendix

Bruner JS (1971) The Relevance of Education. Norton, New York.

Fransella F, Bannister D (1986) Inquiring Man: The Theory of Personal Constructs, 2nd ed. Routledge, London.

Pais A (1982) 'Subtle is the Lord...' The Science and Life of Albert Einstein. Oxford University Press, Oxford.

Quine W van O (1961) From a Logical Point of View, 2nd ed. Harvard University Press, Cambridge, MA. (See especially Essay II, Section 6.)

Toulmin SE (1972) Human Understanding: The Collective Use and Evolution of Concepts. Princeton University Press, Princeton, NJ.

Name Index

Subject Index